AF586919

Control Systems

A Comprehensive Lab Manual

Control Systems

A Comprehensive Lab Manual

Ch. Chengaiah

Professor
Department of Electrical and Electronics Engineering
Sri Venkateswara University College of Engineering
Tirupati.

G. V. Marutheswar

Professor
Department of Electrical and Electronics Engineering
Sri Venkateswara University College of Engineering

BS Publications
A unit of **BSP Books Pvt. Ltd.**
4-4-309/316, Giriraj Lane, Sultan Bazar,
Hyderabad - 500 095
Phone : 040 - 23445605, 23445688

Control Systems : A Comprehensive Lab Manual
by Ch. Chengaiah and G. V. Marutheswar

Published by

BS Publications
A unit of **BSP Books Pvt. Ltd.**
4-4-309/316, Giriraj Lane, Sultan Bazar,
Hyderabad - 500 095
Phone : 040 - 23445605, 23445688
e-mail : info@bspbooks.net
www.bspbooks.net

ISBN: 978-93-86819-79-6 (Hardback)

Contents

Chapter 1

Introduction to Matlab

1.1 What is MATLAB? 1

1.2 How to Start MATLAB 1

1.3 MATLAB Environment 1

1.4 Useful Functions and Operations in MATLAB 2

1.5 Obtaining Help on MATLAB Commands 2

1.6 Variables in MATLAB 3

1.7 Vectors and Matrices in MATLAB 3

1.8 The Dot Allows us to do Operations Element Wise 5

1.9 How to draw a plot with MATLAB 5

Chapter 2

Familiarization with MATLAB Control System Toolbox

2.1 Familiarization with MATLAB Control System Toolbox 7

***Review Questions* 12**

Chapter 3

Transfer Functions

3.1 Introduction 13

3.2 Procedure for Determining the Transfer Function of a Control System 13

3.3 Linear Time Invariant System 14

3.4 Block Diagram Reduction Rules 14

***Review Questions* 16**

***Experiments* 16**

3.1 Transfer Function from Zeros and Poles 16

3.2 Transfer Function to Pole Zero Conversion and Vice Versa 18
3.3 Transfer Function of Different Configurations 21
3.4 Block Diagram Reduction 24
Exercise Problems **........... 26**

Chapter 4

Time Domain Analysis

4.1 Introduction 27
4.2 Transient and Steady State Response 27
4.3 Input Test signals 29
4.4 Impulse Signal 29
4.5 Step Signal 30
4.6 Ramp Signal 30
4.7 Acceleration Signal 31
4.8 Time Response of a First Order Control System 31
4.9 Time Response of a Second Order Control System 32
4.10 Time Response of Second Order Control System 36
4.11 Time Response of Second Order Control System 37
4.12 Step Response from a Given Transfer Function 40
4.13 Impulse Response from a given Transfer Function 42
4.14 Steady State Error Constants 44
Exercise Problems **........... 48**

Chapter 5

Stability Analysis

5.1 Introduction 49
5.2 Definitions of Stability 49
5.3 Absolute and Relative Stability 49
5.4 Zero Input Stability 50
5.5 Asymptotic Stability 50
5.6 Marginal Stability 50

5.7 R-H Criteria 51
5.8 Root Locus 51
5.9 Nyquist Criteria 52
5.10 Bode Plot 53
5.11 Root Locus from a Transfer Function 54
5.12 Bode Plot from a Transfer Function 56
5.13 Nyquist Plot from a Transfer Function 59
***Exercise Problems* 62**

Chapter 6
Compensation and Controllers of Control Systems

6.1 Introduction 63
6.2 Necessary of Compensation 63
6.3 Types of Compensation Techniques 63
6.4 Phase Lead Compensation 65
6.5 Phase Lag Compensation 66
6.6 Phase Lag-Lead Compensation 67
6.7 Introduction 68
6.8 Types of Controllers 68
6.9 Characteristic Analysis of PID Controller 70
6.10 Lag Compensator 71
6.11 Lead Compensator 76
6.12 Lag- Lead Compensator 81
6.13 PID Controller Design 86
***Exercise Problems* 89**

Chapter 7
State Space Analysis in Control Systems

7.1 Introduction 90
7.2 State 90

7.3 State Variable 91
7.4 State Vector 91
7.5 State Space 91
7.6 Stability 92
7.7 Controllability 92
7.8 Observability 93
7.9 Transfer Funtion to State Space Model and Vice Versa 93
7.10 Controllability 96
7.11 Observability 98
7.12 Step Response of a State Space Model 100
7.13 Impulse Response of a State Space Model 103
***Exercise Problems* 105**

Chapter 8

Simulink Model

8.1 Introduction to SIMULINK 107
8.2 Key Features 107
8.3 Building the Model 107
8.4 Selecting Blocks 109
8.5 Building and Editing the Model 108
8.6 Navigating through the Model Hierarchy 108
8.7 Managing Signals and Parameters 110
8.8 Simulink helps you Determine the following Signal and Parameter Attributes 110
8.9 Simulating the Model 110
8.10 Choosing a Solver 109
8.11 Running the Simulation 110
8.12 Analysing Simulation Results 110
8.13 Viewing Simulation Results 110

8.14 Debugging the Simulation 111

***Exercise Problems* 114**

Preface

This book is written primarily as a lab manual of Control Systems for B.Tech Electrical and Electronics Engineering students of all Universities across the nation. It is intended for students in Electrical and other engineering disciplines like Electronics and Communication Engineering, Instrumentation and Control Systems Engineering and Process Instrumentation Engineering etc. as well as being useful as a reference for lab instructors and self-study guide for the professionals dealing in control system area.

This world is living a fast and competitive world and hi-tech technology, this book is designed to know the awareness of fundamentals of MATLAB, how to analyse the systems in MATLAB//SIMULINK environment. The organization and details of the material in this book enables maximum flexibility for the teachers/instructors to select the topics to include in various courses of engineering curriculum.

The book Emphasis is given to an improved appreciation of the operational characteristics of the field of control systems, on the basis of basic every mathematical step. Almost every key concept is illustrated through the use of model graphs that are worked out in detail to enforce the student's understanding.

First of all I owe a debt of gratitude to almighty god to bring a shape of this book in a practical manner.

I would like to thank my teacher Prof. R. V. S. Satyanarayana of Electronics and Communication engineering who encouraged constantly and for his suggestions towards the preparation of this book.

I would like to thank my Family and research scholars who involved and helped me in preparation of this book.

The author will gratefully acknowledge constructive criticism from both students and teachers/instructors for further improvement of this book.

With gratitude

Prof. Ch. Chengaiah
Prof. G. V. Marutheswar

1 Introduction to Matlab

1.1 What is MATLAB?

MATLAB is widely used in all areas of applied mathematics, in education and research at universities, and in the industry. MATLAB stands for Matrix Laboratory and the software is built up around vectors and matrices. This makes the software particularly useful for linear algebra but MATLAB is also a great tool for solving algebraic and differential equations and for numerical integration. It has powerful graphic tools and can produce nice pictures in both 2D and 3D. It is also a programming language, and is one of the easiest programming languages for writing mathematical programs. It also has some tool boxes useful for signal processing, image processing, optimization, etc. MATLAB is available on a number of computing environments: MS windows, DOS, UNIX workstations, Macintosh etc.

1.2 How to Start MATLAB

Mac: Double-click on the icon for MATLAB.

PC: Choose the submenu "Programs" from the "Start" menu. From the "Programs" menu, open the "MATLAB" submenu. From the "MATLAB" submenu, choose "MATLAB".

Unix: At the prompt, type `matlab`.

You can quit MATLAB by typing `exit` in the command window.

1.3 MATLAB Environment

The MATLAB environment (on most computer systems) consists of menus, buttons and a writing area similar to an ordinary word processor. There are plenty of help functions that you are encouraged to use. The writing area that you will see when you start MATLAB, is called the *command window*. In this window you give the commands to MATLAB. For example, when you want to run a program you have written for MATLAB you start the program in the command window by typing its name at the prompt. The command window is also useful if you just want to use MATLAB as a scientific calculator or as a graphing tool. If you write longer programs, you will find it more convenient to write the program code in a separate window, and then run it in the command window.

In the command window you will see a prompt that looks like >>. You type your commands immediately after this prompt. Once you have typed the command you wish MATLAB to perform, press <enter>. If you want to *interrupt a command* that MATLAB is running, type <ctrl> + <c>.

The commands you type in the command window are stored by MATLAB and can be viewed in the *Command History* window. To repeat a command you have already used, you can simply double-click on the command in the history window, or use the <up arrow> at the command prompt to iterate through the commands you have used until you reach the command you desire to repeat.

1.4 Useful Functions and Operations in MATLAB

Using MATLAB as a calculator is easy.

Example: Compute 5 $\sin(2.5^{3-pi})+1/75$. In MATLAB this is done by simply typing

```
5*sin(2.5^(3-pi))+1/75
```

at the prompt. Be careful with parentheses and don't forget to type `*` whenever you multiply.

Note: MATLAB is *case sensitive.* This means that MATLAB knows a difference between letters written as lower and upper case letters. For example, MATLAB will understand `sin(2)` but will not understand `Sin(2)`.

1.5 Obtaining Help on MATLAB Commands

To obtain help on any of the MATLAB commands, you simply need to type

```
help <command>
```

at the command prompt. For example, to obtain help on the `gamma` function, we type at the command prompt:

```
help gamma
```

Try this now. You may also get help about commands using the "Help Desk", which can be accessed by selecting the MATLAB Help option under the Help menu.

Note: The description in MATLAB returns about the command you requested help on contains the command name in ALL CAPS. This does not mean that you use this command by typing it in ALL CAPS. In MATLAB, you almost always use all lower case letters when using a command.

1.6 Variables in MATLAB

We can easily define our own variables in MATLAB. Let's say we need to use the value of 3.5 sin (2.9) repeatedly. Instead of typing `3.5*sin(2.9)` over and over again, we can denote this variable as x by typing the following:

```
x = 3.5*sin(2.9)
```

Often, we may not want to have the result of a calculation printed-out to the command window. To supress this output, we put a semi-colon at the end of the command; MATLAB still performs the command in "the background". If you defined x as above, now type as shown below and observe.

```
y = 2 * x;
y
```

In many cases we want to know what variables we have declared. We can do this by typing `whos`. Alternatively, we can view the values by opening the "Workspace" window. This is done by selecting the Workspace option from the View menu. If you want to erase all variables from the MATLAB memory, type `clear`. To erase a specific variable, say x, type `clear x`. To clear two specific variables, say x and y, type `clear x y`, that is separate the different variables with a space. Variables can also be cleared by selecting them in the Workspace window and selecting the delete option.

1.7 Vectors and Matrices in MATLAB

We create a vector in MATLAB by putting the elements within [ ] brackets.

Example: `x = [1 2 3 4 5 6 7 8 9 10]`

We can also create this vector by typing `x = 1:10.` The vector (`1 1.1 1.2 1.3 1.4 1.5`) can be created by typing

`x =[1 1.1 1.2 1.3 1.4 1.5 ]` or by typing

`x = 1:0.1:1.5.`

Matrices can be created according to the following example. The matrix $A = \begin{bmatrix} 1 & 2 & 3 \\ 4 & 5 & 6 \\ 7 & 8 & 9 \end{bmatrix}$ is created by typing

```
A = [1 2 3 ; 4 5 6; 7 8 9],
```

i.e., rows are separated with semi-colons. If we want to use a specific element in a vector or a matrix, study the following example:

Example:

```
x=[10 20 30]
A=[ 1 2 3 ; 4 5 6 ; 7 8 9]
x(2) represents second element of X vector
A(3,1) represents third row first elememt of Matrix
A
```

Here we extracted the second element of the vector by typing the variable and the position within parantheses. The same principle holds for matrices; the first number specifies the row of the matrix, and the second number specifies the column of the matrix.

If the matrices (or vectors which are special cases of a matrices) are of the same dimensions then matrix addition, matrix subtraction and scalar multiplication works just like we are used to.

Example: Type

```
x =[1 2 3]
y =[4 5 6]
a = 2
x + y
x - y
a × x
```

If want to apply an operation such as squaring each element in a matrix we have to use a dot. before the operation we wish to apply. Type the following commands in MATLAB.

```
x = 1:10
x .^2
A = [1 2 3 ; 4 5 6 ; 7 8 9 ]
A.^2
A^2
```

1.8 The Dot Allows us to do Operations Element Wise

All built-in functions such as `sin`, `cos`, `exp` and so on automatically act element wise on a matrix. Type

```
Y =[0 1/4 1/2 3/4 1]
y = pi*y
sin(y)
```

1.9 How to draw a plot with MATLAB

There are different ways of plotting in MATLAB. The following two techniques, illustrated by examples, are probably the most useful ones.

Example 1: Plot $\sin(x^2)$ on the interval [–5, 5]. To do this, type the following:

```
x =-5:0.01:5;
y = sin(x.^2);
plot(x,y)
```

Example 2: Plot exp(sin(x)) on the interval $[-\pi, \pi]$. To do this, type the following:

```
x=linspace(-pi,pi,101);
y = exp(sin(x));
plot(x,y)
```

The command `linspace` creates a vector of 101 equally spaced values between $-\pi$ and π (inclusive).

Occasionally, we need to plot values that vary quite differently in magnitude. In this case, the regular `plot` command fails to give us an adequate graphical picture of our data. Instead, we need a command that plots values on a log scale. MATLAB has 3 such commands: `loglog`, `semilogx`, and `semilogy`. Use the `help` command to see a description of each function. As an example of where we may want to use one of these plotting routines, consider the following problem:

Example 3: Plot $x^{5/2}$ for $x = 10^{-5}$ to 10^5. To do this, type the following:

```
x=logspace(-5,5,101);
y=x.^(5/2);
plot(x,y)
```

Now type the following command:

```
loglog(x,y)
```

The command `logspace` is similar to `linspace`, however it creates a vector of 101 points logarithmically equally distributed between 10^{-5} and 10^{5}.

- **Options**

  Plot, loglog, semilogx, semilogy, bar, hist, stem, polar, contour.

- **Graphic Commands**

  Title, xlabel, ylabel, text, grid, axis, clf, ginput, gtext.

- **Line types**

  Dotted .. dashes-- dashdot-.

- **Point Type**

  Point. Plus+ star* circle o xmark x

- **Color**

  Red r green g blue b yellow y magenta m

  Cyan c white w black k

- **Multiple plots**

  Subplot(mnp) options:221 211 121 ....

- **Auto scaling**
- **Zooming** zoom on

2 Familiarization with MATLAB Control System Toolbox

2.1 Familiarization with MATLAB Control System Toolbox

Program-1

Consider a plant transfer function $G(s) = \dfrac{s^3 + 5s^2 + 4s + 6}{4s^3 + 7s^2 + 12s + 9}$

- To enter a transfer function
- Transfer function to pole-zero conversion
- To draw the pole zero plot
- Pole-zero to transfer function
- To find the partial fraction expansion of the transfer function
- r, p, k to transfer function
- Transfer function to state space conversion
- State space to transfer function
- To find the eigen value of the function
- To find the eigen value of the system
- MIMO system state space to transfer function

```
% Program-1
%(1) To enter a Transfer function
num1= [1 5 4 6];
den1= [4 7 12 9];
sys1 = tf(num 1, den 1)
%(2) Transfer function to pole- zero conversion
[z,p,k] = tf2zp(num1,den1)
%(3) To draw the pole zero plot
pzmap(num1,den1)
% (4) zp2tf to zero pole to transfer function
[num,den] = zp2tf(z,p,k)
```

% (5) To find the partial fraction expansion of the transfer function.

[r,p,k] = residue(num1,den1)

%(6) r,p,k to transfer function

[num2,den2] = residue(r,p,k)

%(7) Transfer function to state space conversion

[A,B,C,D] = tf2ss(num1,den1)

%(8) State space to transfer function

[num3,den3] = ss2tf(A,B,C,D)

%(9) obtain the various transfer function of a MIMO SYSTEM

A = [0 1; –25 –4];

B = [1 1;0 1];

C = [1 0;0 1];

D = [0 0;0 0];

[num1,den1] = ss2tf(A,B,C,D,1);

[num2,den2] = ss2tf(A,B,C,D,1)

%(10) To find the eigen value of the function

eig(A)

Program-2

$G_c(s) = \frac{S+1}{S+2}$ $P(s) = \frac{1}{500S^2}$ are the controller and the plant transfer functions obtain the overall transfer function of a system connected in

(i) Cascade
(ii) parallel
(iii) feedback

% (i) Cascaded systems

Numg=[1];deng=[500 0 0];

numh=[1 1];denh=[1 2];

[numc,denc]=series(numg,deng,numh,denh)

printsys(numc,denc)

% (ii) Parallel systems

[nump,denp]=parallel(numg,deng,numh,denh)

printsys(nump,denp)

% (iii) Negative unity feedback systems

```
[nf,df]=cloop(numc,denc,-1)
printsys(nf,df)
```

% (iv) Positive unity feedback systems

```
[nfp,dfp]=cloop(numc,denc,1)
printsys(nfp,dfp)
```

% (v) Feedback systems

```
[nums,dens]=feedback(numg,deng,numh,denh,-1)
printsys(nums,dens)
```

Program-3

Consider the transfer function G(s)

$$=\frac{6S^2+1}{S^3+3S^2+3S+1} \text{ and } H(s)=\frac{(S+1)(S+2)}{(S+2i)(S-2i)(S+3)}$$

Compute the following (i) compute the poles and zeros of G(s) (ii) write H(s) as a function of two polynomials (iii) Divide G(s)/H(s), also its pole zero plot.

% (i) Compute the poles and zeros of G(s)

```
numg=[6 0 1];deng=[1 3 3 1];
Z=roots(numg)
P=roots(deng)
```

%(ii) Write H(s) as a function of two polynomials

```
n1=[1 1];
n2=[1 2];
d1=[1 2*i];
d2=[1 -2*i];
d3=[1 3];
numh=conv(n1,n2);
denh=conv(conv(d1,d2),d3);
printsys(numh,denh)
```

%(iii) Divide G(s)/H(s), also find its pole zero plot.

```
num=conv(numg,denh);
den=conv(deng,numh);
printsys(num,den)
pzmap(num,den)
```

Program-4

Using Matlab command find the magnitude and phase angle with Transfer function $G(s) = \frac{5}{S+2}$ at w = 3rad/sec

```
gnum=[5];
gden=[1 2];
S=3*j;
gj3=polyval(gnum,S)/polyval(gden,S);
maggj3=abs(gj3)
phasegj3=angle(gj3)*(180/pi)
```

Program-5

Consider a second order system with standard second order transfer functions $G(s) = \frac{w^2{}_n}{S^2 + 2\zeta w_n S + w^2{}_n}$, by defining the normalized frequency

$\omega_{v} = \frac{\omega}{\omega_v}$ plot the frequency response for various values of ζ =0.25,0.5,0.707,1

```
w = 0:0.05:3;
z = [0.25 0.5 0.707 1]
for k = 1:4
gnum = [0 0 1];gden = [1 2*z(k) 1]
gjomega = freqs(gnum,gden,w);
gmag=abs(gjomega);
plot(w,gmag)
title('frequency response of G(s)')
xlabel('omega');
```

```
ylabel(' G|jw| ');
grid;hold on
end
```

Exercise 1

Repeat the program - 1 for the following plant transfer functions

(i) $G(s) = \dfrac{10}{S(S+1)(S+5)}$

(ii) $G(s) = \dfrac{2S(S+1)(S+9)}{(S+4)(S+5)(S+7)}$

(iii) $G(s) = \dfrac{S^2}{(S+3)(S+8)(S+12)}$

Exercise 2

Repeat the program –2 for the following transfer functions

$$G_c(s) = \frac{1}{S^2+2S+5} \quad P(s) = \frac{125}{S^2+50S+125}$$

Exercise 3

Repeat the program -3 for the following transfer functions

$$G(s) = \frac{S^2+2S+1}{(S+4(S+7)(S+15)} \text{ and } H(s) = \frac{S+1}{S^2+S+7}$$

Exercise 4

Consider the polynomials P(s)=S+2S+1 and q(s)=S+1, compute the following using MATLAB

(i) P(s)q(s)

(ii) Poles and zeros of $G(s) = \dfrac{p(s)}{q(s)}$

(iii) P(-1)

Exercise 5

Obtain the eigen value, eigen vector, rank, determinant and inverse of a matrix

$$A = \begin{bmatrix} 0 & 1 & 0 \\ 0 & 0 & 1 \\ -6 & -11 & -6 \end{bmatrix}$$

Review Questions

1. Give the command to find Pole-Zero map?
2. What is the command to find transfer function?
3. What is the command to find partial fraction?
4. Give the command to find Pole-Zero map
5. What is the command to transfer state space to transfer function?

3 Transfer Functions

3.1 Introduction

For any control system there exist an input termed as excitation which operates through a transfer operation termed as transfer function and produces an effect resulting in response termed as controlled variables. Therefore the transfer function is nothing but mathematical equivalent of system. The order of transfer function represents the number of storage elements or number of time constant. The transfer function of a linear time invariant system is defined as the ratio of Laplace transform of output to the Laplace transform of input with all initial conditions being assumed to be zero. The relation between the output and input is represented by the block diagram for a control system as shown in Fig.2.1 and corresponding mathematical expression is:

$$G(s) = \frac{L[\text{Output}]}{L[\text{Input}]} = \frac{C(S)}{R(s)}$$

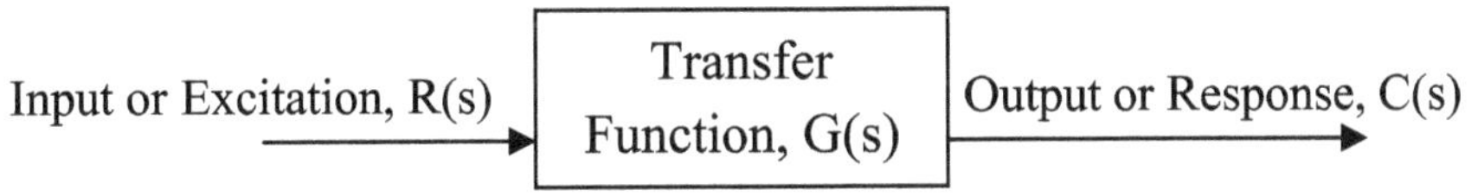

Input or Excitation, R(s)

Transfer Function, G(s)

Output or Response, C(s)

Fig.3.1 Block diagram of control system

3.2 Procedure for Determining the Transfer Function of a Control System

The following steps give a procedure for determining the transfer function of a control system.

1. Formulate the equations for the system.
2. Take the Laplace transform of the system equations assuming zero initial conditions.
3. Specify the system output and the input.
4. Take the ratio of Laplace transform of output to the Laplace transform of input

The ratio obtained in step (4) is the required transfer function.

3.3 Linear Time Invariant System

- LTI system is RLC network, because the RLC components give the linear transfer characteristics and RLC component values are not changing w.r.t time.
- In the T.F analysis, the initial conditions must be zero to get the linear transfer characterises.

3.4 Block Diagram Reduction Rules

For a complex control system it is not convenient to derive a complete transfer function. Therefore the transfer function of each element of a control system is represented by a block diagram and the concerned symbol mentioned in the block represents the transfer function of the element. This symbolic representation in a short form is a pictorial representation relating the output and input of control system based on cause and effect approach.

In order to obtain the overall transfer function a procedure called block diagram reduction technique is followed.

Rule-1 Summation of summing points

$Y = X_1 - X_2 - X_3$

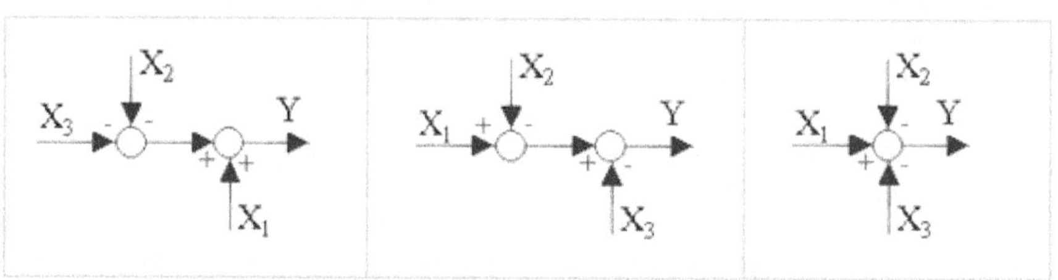

Rule-2 Associative and commutative properties.

$Y = G_1 G_2 X = G_2 G_1 X$

Rule-3 Distributive property

$$Y = G_1(X_1 - X_2) = G_1X_1 - G_1X_2$$

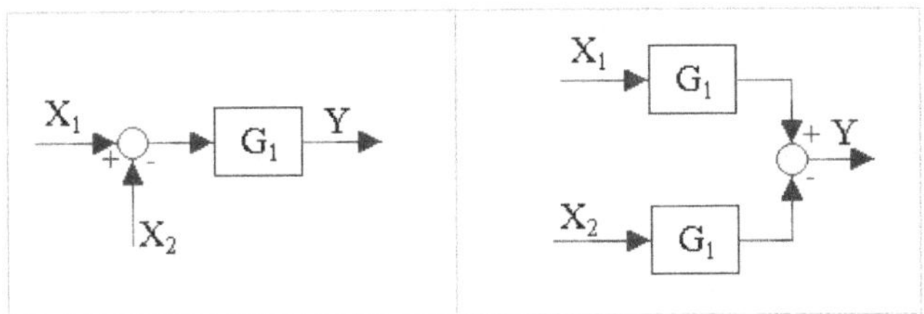

Rule-4 Blocks in parallel

$$Y = X(G_1 + G_2) = G_1X + G_2X$$

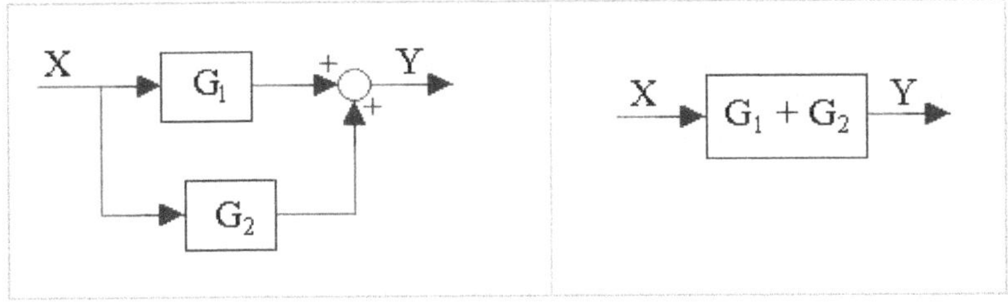

Rule-5 Positive feedback loop.

$$Y = G_1X + G_2G_1Y = \frac{G_1}{1 - G_1G_2}X$$

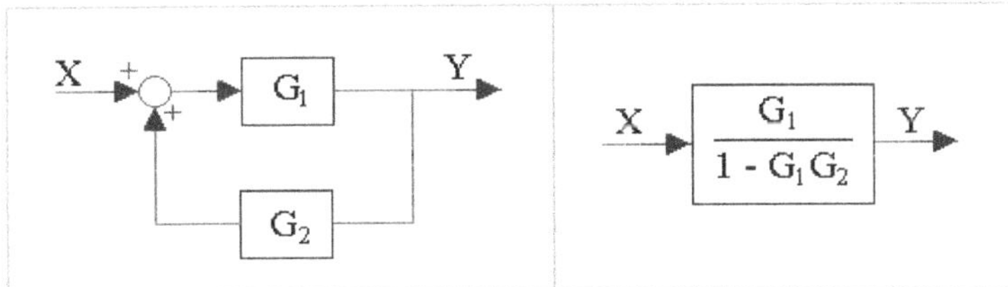

$$\frac{\text{Output}}{\text{Input}} = \frac{\text{feed forward transfer function}}{1 - \text{feedforward x feedback}}$$

Rule-6 Negative feedback loop.

$$Y = G_1X - G_2G_1Y = \frac{G_1}{1+G_1G_2}X$$

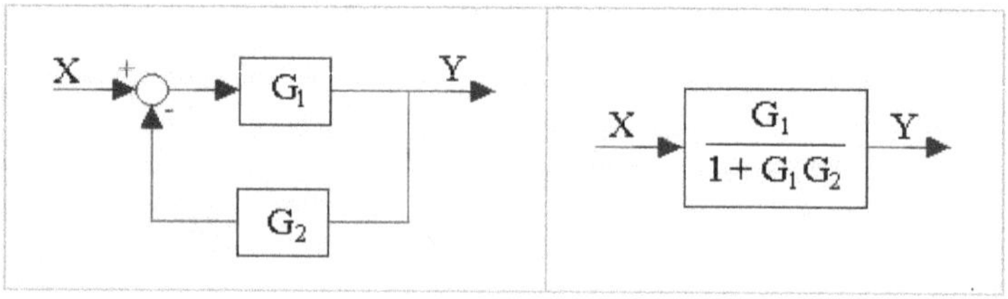

$$\frac{\text{Output}}{\text{Input}} = \frac{\text{feed forward transfer function}}{1 + \text{feedforward x feedback}}$$

Review Questions

1. What do you mean by a control system?
2. When do you say that mathematical model is linear?
3. When do you say that the model is linear time variant?
4. When do you say that the model is linear time invariant?
5. What do you mean by feedback?
6. Why negative feedback if preferred in control systems?

EXPERIMENTS

3.1 Transfer Function from Zeros and Poles

Aim: To obtain a transfer function from given poles and zeroes using MATLAB

Apparatus: Personal Computer (PC)

MATLAB Software

Theory:

A transfer function is also known as the network function is a mathematical representation, in terms of spatial or temporal frequency, of the relation between the input and output of a (linear time invariant) system. The transfer function is the ratio of the output Laplace Transform to the input Laplace

Transform assuming zero initial conditions. Many important characteristics of dynamic or control systems can be determined from the transfer function.

The transfer function is commonly used in the analysis of single-input single-output electronic system, for instance. It is mainly used in signal processing, communication theory, and control theory. The term is often used exclusively to refer to linear time-invariant systems (LTI). In its simplest form for continuous time input signal x(t) and output y(t), the transfer function is the linear mapping of the Laplace transform of the input, X(s), to the output Y(s).

Zeros are the value(s) for z where the numerator of the transfer function equals zero. The complex frequencies that make the overall gain of the filter transfer function zero. Poles are the value(s) for z where the denominator of the transfer function equals zero. The complex frequencies that make the overall gain of the filter transfer function infinite.

The general procedure to find the transfer function of a linear differential equation from input to output is to take the Laplace Transforms of both sides assuming zero conditions, and to solve for the ratio of the output Laplace over the input Laplace.

MATLAB Program:

```
Clc;
Clear all;
z=input('enter zeroes')
p=input('enter poles')
k=input('enter gain')
[num,den]=zp2tf(z,p,k)
tf(num,den)
```

Example:

Construct the transfer function for the following zeros, poles and gain of the system.

```
z = -1.0000 + 1.4142i
    -1.0000 - 1.4142i
p = -1.2653 + 0.0000i
    -0.0340 + 1.2568i
    -0.0340 - 1.2568i
k =  0.3333
```

Output:

System Transfer Function:

$$\frac{S^2 + 2s + 3}{3\,S^3 + 4\,S^2 + 5\,s + 6}$$

Continuous-time transfer function.

Result:

Viva -Voce:

1. Define zero and pole.
2. Define Type and order
3. Explain significance of Transfer function.
4. Mention drawbacks of transfer function.
5. What do you mean by zero initial conditions?

3.2 Transfer Function to Pole Zero Conversion and Vice Versa

Aim: To convert the given transfer function into pole-zero conversion and vice versa.

Apparatus: Personal Computer (PC)

MATLAB software

Theory:

The pole-zero and transfer function representations of a system are tightly linked. For example consider the transfer function:

$$H(S) = \frac{b_0S^2 + b_1S + b_2}{a_0S^3 + a_1S^2 + a_2S + a_3}$$

If we rewrite this in a standard form such that the highest order term of the numerator and denominator are unity (the reason for this is explained below).

$$H(S) = \frac{b_0}{a_0} \frac{S^2 + \frac{b_1}{b_0} S + \frac{b_2}{b_0}}{S^3 + \frac{a_1}{a_0} S^2 + \frac{a_2}{a_0} S + \frac{a_3}{a_0}}$$

This is just a constant term (b_0/a_0) multiplied by a ratio of polynomials which can be factored.

$$H(S) = K \frac{(s - z_1)(s - z_2)}{(s - p_1)(s - p_2)(s - p_3)}$$

In this equation the constant $k=b_0/a_0$. The z_i terms are the zeros of the transfer function; as $s \to z_i$ the numerator polynomial goes to zero, so the transfer function also goes to zero. The p_i terms are the poles of the transfer function; as $s \to p_i$ the denominator polynomial is zero, so the transfer function goes to infinity. In the general case of a transfer function with an mth order numerator and an nth order denominator, the transfer function can be represented as:

$$H(S) = K \frac{\prod_{i=1}^{m} (s - z_i)}{\prod_{i=1}^{n} (s - p_i)}$$

The pole-zero representation consists of the poles (p_i), the zeros (z_i) and the gain term (k). Often the gain term is not given as part of the representation. The nature of the behaviour of the system is given by the poles and zeros, the gain term only determines the magnitude of the response. In many cases a plot is made of the s-plane that shows the locations of the poles and zeros, and the gain term (k) is not shown.

MATLAB Program:

```
clc;
clear all;
num=input('Enter the coefficients of numerator')
den=input('Enter the coefficients of denominator')
sys=tf(num,den)
disp('Transfer function to pole-zero conversion')
disp('zeros,poles,gain of transfer function');
[z p k]=tf2zp(num,den)
```

```
disp('To draw the pole zero plot')
pzmap(num,den)
disp('pole zero to transfer function')
[num1,den1]=zp2tf(z,p,k)
```

Example:

Find poles zero map by finding poles and zeros the given transfer function

$$T(s) = \frac{S^3 + 5S^2 + 4S + 6}{4S^3 + 7S^2 + 12S + 9}$$

Output:

Zeros, Poles and Gain of Tansfer function

z = –4.4009

–0.2996 + 1.1286i

–0.2996 – 1.1286i

p = –0.3750 + 1.4524i

–0.3750 – 1.4524i

–1.0000

k = 0.2500

Pole-Zero plot:

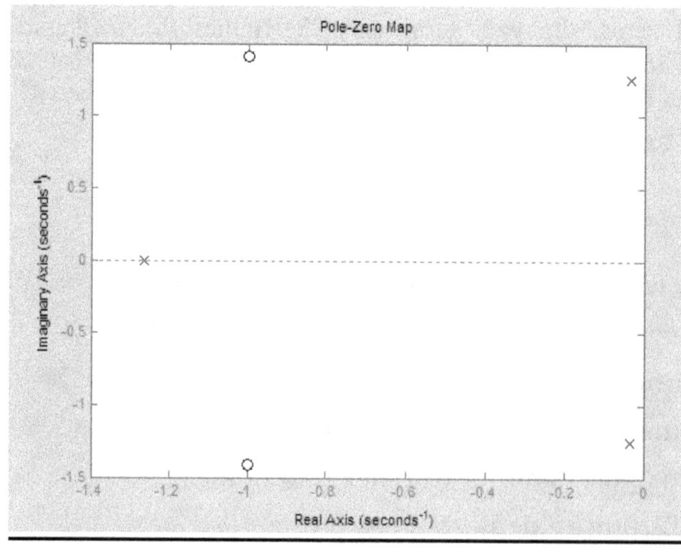

Result:

Viva-Voce:

1. Define transfer function?
2. Why initial condition is taken as zero in transfer function?
3. Explain the significance of pole-zero placement.
4. Define BIBO stability?
5. State the principle of superposition.

3.3 Transfer Function of Different Configurations

Aim: To obtain the transfer function of different configurations.

Apparatus: Personal Computer (PC)

MATLAB software

Theory:

The block diagram modeling may provide control engineers with a better understanding of the composition and interconnection of the components of a system. It can be used, together with transfer functions, to describe the cause- effect relationships throughout the system. The common elements in block diagrams of most control systems include:

- Comparators
- Blocks representing individual component transfer functions, including:
    - Reference sensor (or input sensor)
    - Output sensor
    - Actuator
    - Controller
    - Plant (the component whose variables are to be controlled)
- Input or reference signals
- Output signals
- Disturbance signal
- Feedback loop

Some of the block diagram reduction Rules are given below:

1. Two blocks connected in series

2. Two blocks connected in parallel

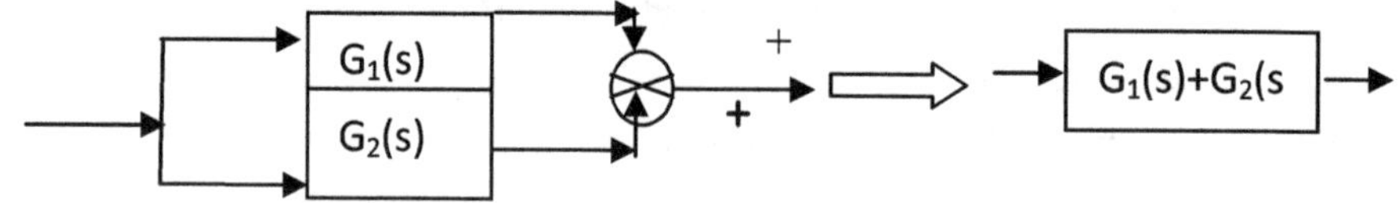

3. Two blocks connected in Feedback.

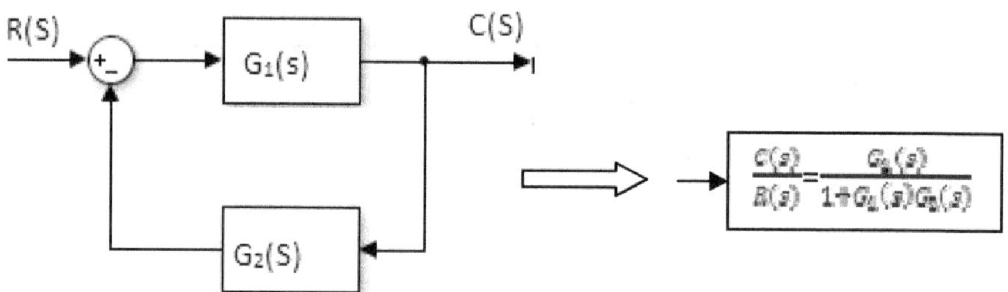

MATLAB Program:

```
clc;
clear all;
num=input('Enter the cofficients of numerator');
den=input('Enter the cofficients of denominator');
g=tf(num,den)
num1=input('Enter the cofficients of numerator');
den1=input('Enter the cofficients of denominator');
gc=tf(num1,den1)
disp('Transfer function when connected in cascade form')
c=g*gc
disp('Transfer function when connected in parallel form')
p=g+gc
```

```
disp('Transfer function when connected in feedback')
c=feedback(gc,[g])
```

Example:

Obtain the transfer function of following two systems in different configurations:

Transfer function1: $\frac{S+1}{S+2}$

Transfer function2: $\frac{1}{500\,s^2}$

Output:

Enter the cofficients of numerator [1 1]

Enter the cofficients of denominator [1 2]

Transfer function: $\frac{s+1}{s+2}$

Enter the cofficients of numerator [0 0 1]

Enter the cofficients of denominator [500 0 0]

Transfer function: $\frac{1}{500\,s^2}$

Transfer function when connected in cascade form:

$$\frac{s+1}{500\,s^3+1000\,s^2}$$

Transfer function when connected in parallel form:

$$\frac{500\,s^3\;500\,s^2+s+2}{500\,s^3+1000\,s^2}$$

Transfer function when connected in feedback:

$$\frac{s+2}{500\,s^3+1000\,s^2+s+1}$$

Result:

Viva-voce:

1. What do you mean by a block?
2. What is G(s) and H(S)?
3. Name the components of block diagram?
4. What are the can be performed by sensing devices?
5. Block diagrams that diagrams can be used to model non-linear systems?

3.4 Block Diagram Reduction

Aim: To obtain the transfer function for the following block diagram using block diagram reduction techniques.

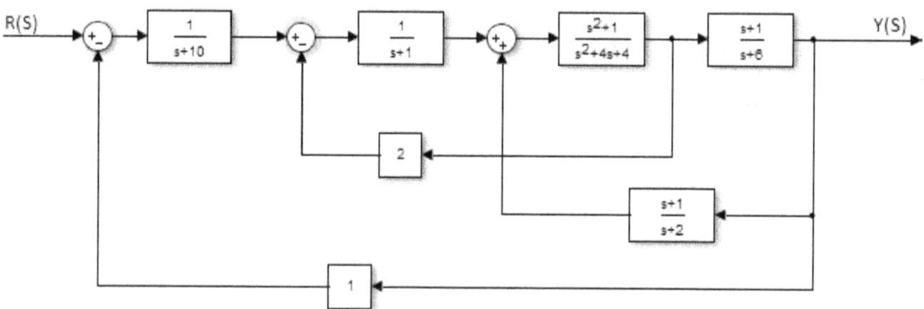

Apparatus: Personal Computer(PC)

MATLAB software

MATLAB Program:

```
disp('Bolck diagram reduction- obtain the closed
loop transfer function')
numg1=[0 1];
deng1=[1 10];
numg2=[0 1];
deng2=[1 1];
numg3=[1 0 1];
deng3=[1 4 4];
numg4=[1 1];
deng4=[1 6];
```

```
numh1=[1 1];
denh1=[1 2];
numh2=[2];
denh2=[1];
numh3=[1];
denh3=[1];
disp('Moving a pick off point beyond G4')
n1=conv(numh2,deng4);
d1=conv(denh2,numg4);
[n2a,d2a]=series(numg3,deng3,numg4,deng4);
[n2,d2]=feedback(n2a,d2a,numh1,denh1,1);
[n3a,d3a]=series(n2,d2,numg2,deng2);
[n3,d3]=feedback(n3a,d3a,n1,d1);
[n4,d4]=series(numg1,deng1,n3,d3);
[num,den]=cloop(n4,d4,-1)
```

Output:

```
Bolck diagram reduction- obtain the closed loop
transfer function
Moving a pick off point beyond G4
num =
   0     0     1     4     6     6     5     2
den =
   0    12   205   1066   2517   3128   2196    712
```

Result:

Viva-Vocc:

1. What is the function of the error detector?
2. Mention the merit and demerits of the closed loop system.

3. What is an actuating signal?
4. What do you mean by a forward path gain?
5. What do you mean by loop gain?

Exercise Problems

1. obtain pole-zero map of the following transfer functions.

(a) $T(s) = \dfrac{4}{S^2 + 4S + 16}$

2. $T(s) = \dfrac{5S^2 + 2S + 1}{S(S+2)(S+8)(S+45)}$ Obtain the overall transfer function of the system by block diagram reduction technique

$G_1(s) = \dfrac{1}{S+10}$; $G_2(s) = \dfrac{1}{S+1}$; $G_3(s) = \dfrac{S^2+1}{S^2+4S+4}$;

$G_4(s) = \dfrac{S+1}{S+6}$; $H_1(s) = \dfrac{S+1}{S+2}$

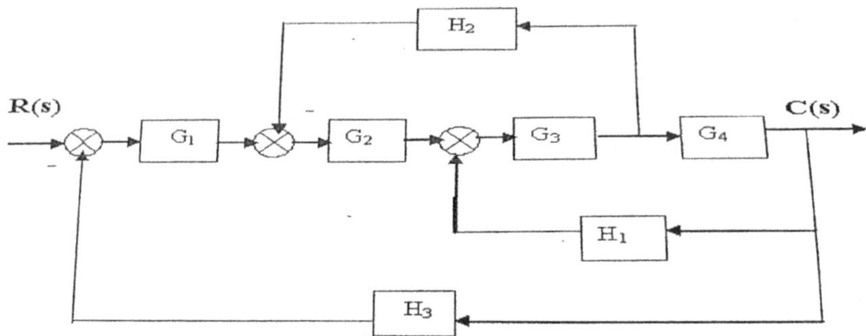

4 Time Domain Analysis

4.1 Introduction

As the input to a control system cannot be assessed beforehand therefore an input test signal is specified for ascertaining comparative time performance of control system. Specified input is one to which the control system is frequently subjected to. Once a control system responds satisfactorily to a test input signal its time response to actual input signal is invariably up to the mark.

Therefore the time response of a control system means as to how a system behaves in accordance with time when a specified test signal is applied. In the initial part of time response of a control system transients appear and during the post transient part steady state is achieved.

The transient and steady state solution are two parts of a differential equation for any physical system. The steady state response of any system gives an idea of the accuracy of the system and if there is deviation between the input and output, the system is said to possess a steady state error, when time tends to infinity. On the other hand, the transient solution of the differential equation gives the nature of variation of the system variables during the transient period. Due to inertia, friction and other energy storing elements in the system, the output may not follow the input at all times during the transient period. During this transient period, the deviation could be sometimes excessive which is not desirable. The system then is said to have got a high overshoot, sometimes excessive which is not desirable. The system then is said to have got a high overshoot. Some systems may not have any overshot at all but they could be very slow to reach the steady level. This is also not desirable in control systems and such systems are said to be sluggish.

4.2 Transient and Steady State Response

In actual practice the steady state period and transient period is identified in terms of time constants of a control system. Thus the time response of a control system is divided into transient response and steady state response for a specified input test signal. The transient response or time response or otherwise called dynamic response of a system gives the variation of the system output variable with respect to time. A simple closed loop block diagram is shown in Fig.4.1 and the corresponding typical time response of a control system as shown in Fig.4.2.

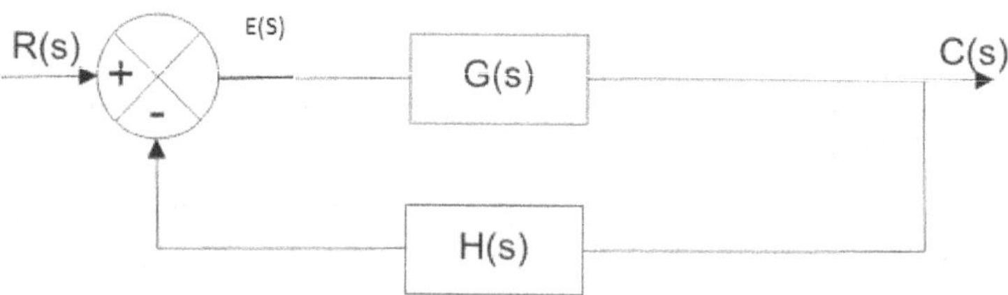

Fig.4.1 Simple closed loop block diagram.

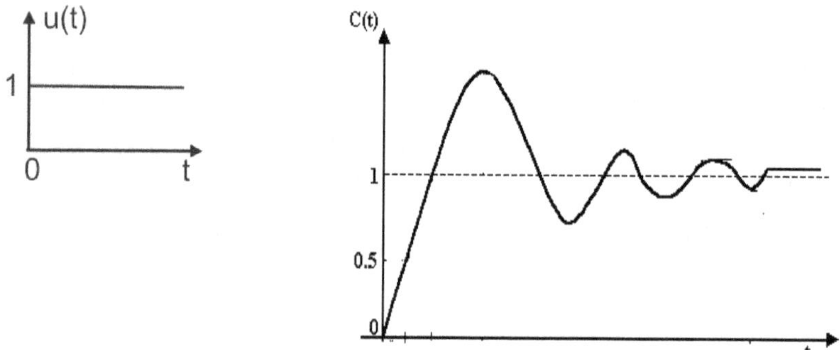

Fig.4.2 Time response of a control system

From Fig.4.1 R(s) is the input variable, C(s) is the output variable, E(s) is the error, G(s) the forward path transfer function and H(s) is the transfer function of the feedback path from fig, the following relationship is easily obtained.

$$E(S) = \frac{R(s)}{1 + GH(s)}$$

$$C(S) = \frac{R(s)G(s)}{[1 + GH(s)]}$$

If the Laplace transform of the input is known, then the transient response of the system variables is easily obtained by taking the Inverse Laplace.

From Fig.4.2 the transient part of time response reveals the nature of response i.e., oscillatory or over damped and also gives an indication above its speed. The steady state part of time response reveals the accuracy of control system. Steady state error is observed if the actual output does not exactly match with the input.

4.3 Input Test signals

System transient response is studied when the nature of input signal applied to the system is known. The commonly used input signals are impulse, step, ramp or velocity, acceleration and sinusoidal signals. Therefore, for transient response study, the first three of the input test signals are usually applied and their corresponding characteristics are described below.

4.4 Impulse Signal

it is also called shock input shown in Fig.4.3 occurs for a small interval of time. Mathematically it is expressed as

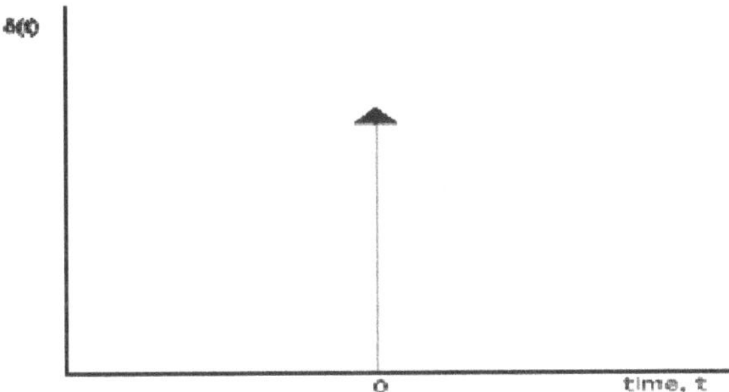

Fig.4.3 Impulse Function

$$r(t) = \delta(t) \text{ for } 0 < t < \frac{1}{H} \text{ Where } H \rightarrow \infty$$

For unit impulse,

$$\delta(t) = 1 \text{ for } t=0$$

$$= 0 \text{ for } t \neq 0$$

The Laplace transform of unit impulse is obtained as

$$R(s) = 1$$

4.5 Step Signal

The time characteristics of a step input is Fig.4.4 mathematically the step input is expressed as

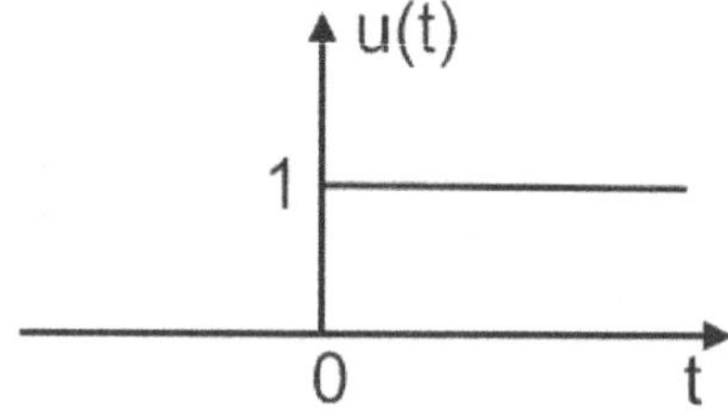

Fig.4.4 Step function

$$r(t) = R \text{ for } t \geq 0$$

the Laplace transform of r(t) is $R(s) = \frac{R}{s}$

For unit step input R=1 and hence $R(s) = \frac{1}{s}$

4.6 Ramp Signal

The ramp or velocity signal is represented in Fig.4.5. It increases linearly with time with a slope of ω_1 rad/sec. The ramp signal is mathematically expressed as,

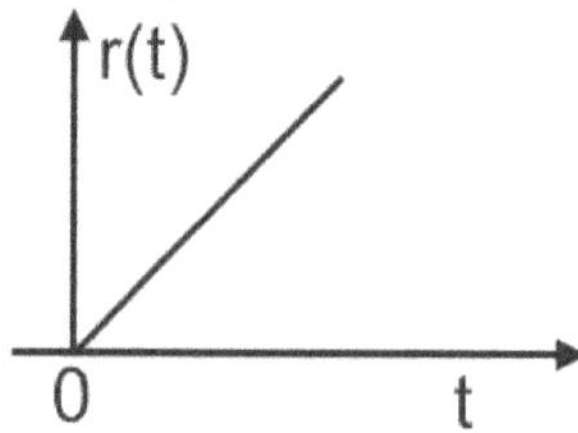

Fig.4.5 Ramp function

$$r(t) = \omega_1 t \quad \text{for } t \geq 0$$

by taking Laplace transform on both sides of the above equation, the L.T of the ramp signal is obtained as

$$R(s) = \frac{\omega i}{s^2}$$

For unit ramp $\omega_1 = 1$ and hence $R(s) = \frac{1}{s^2}$

4.7 Acceleration Signal

The acceleration signal or parabolic signal is shown in Fig.4.6 it is described by the following mathematical expression.

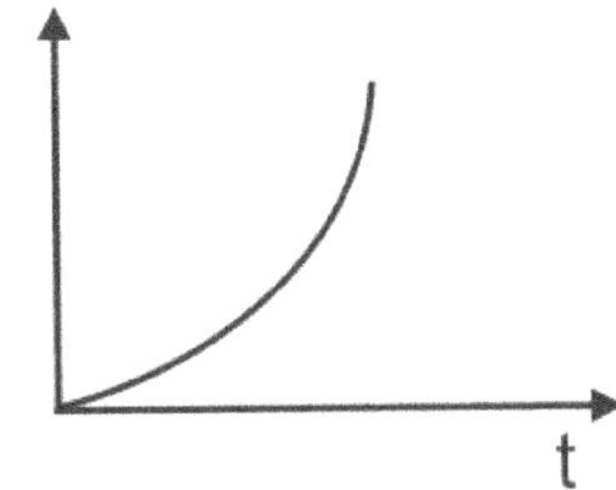

Fig.4.6 Accelerating function

$$r(t) = \frac{At^2}{2} \quad \text{for } t \geq 0$$

The L.T. or the above signal is derived as $R(s) = \frac{A}{s^2}$

4.8 Time Response of a First Order Control System

A first order control system is one wherein highest power of s in the denominator of its transfer function equals to 1. Thus a first order control system is expressed by a transfer function given below:

$$\frac{C(s)}{R(s)} = \frac{1}{1+sT} = G(s)$$

The block diagram representation of the above equation is shown in Fig.4.7

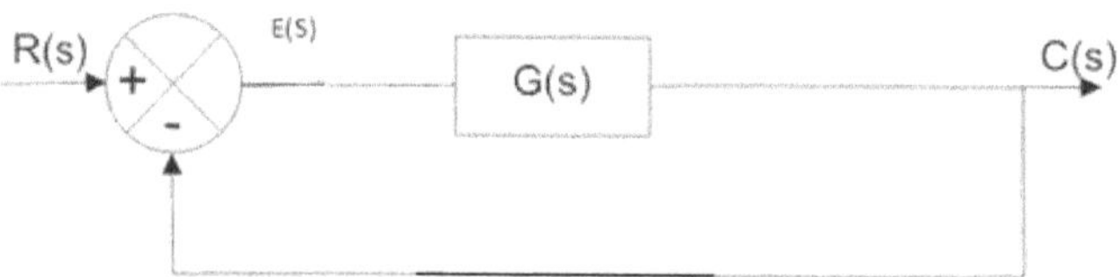

Fig.4.7 Block diagram representation of a first order control system.

The time response of a first order control system subjected to unit step, ramp and impulse are summarised in the Table.4.1

Table.4.1 Response of control system for different inputs

Input function	Time response expression	Observations
Unit ramp $\frac{1}{s^2}$	$C(t)=t-T+T\,e^{-t/T}$	Differentiate ↓ ↑ Integrate
Unit step $\frac{1}{s}$	$C(t)=1-e^{-t/T}$	
Unit impulse 1	$C(t)=\frac{1}{T}e^{-t/T}$	

From Table.4.1 is observed that the step function is first derivative of ramp function and impulse function is first derivative of step function. From the derived time response expression it is concluded that the output time response also follows the same sequence as that of input functions.

4.9 Time Response of a Second Order Control System

A second order control system is one wherein highest power of s in the denominator of its transfer function equals to 2. A general expression of the transfer function of a control system is given by:

$$\frac{C(s)}{R(s)}=\frac{\omega_n^2}{s^2+2\zeta\omega_n s+\omega_n^2}=G(s)$$

The block diagram representation of the above equation is shown in Fig.4.8

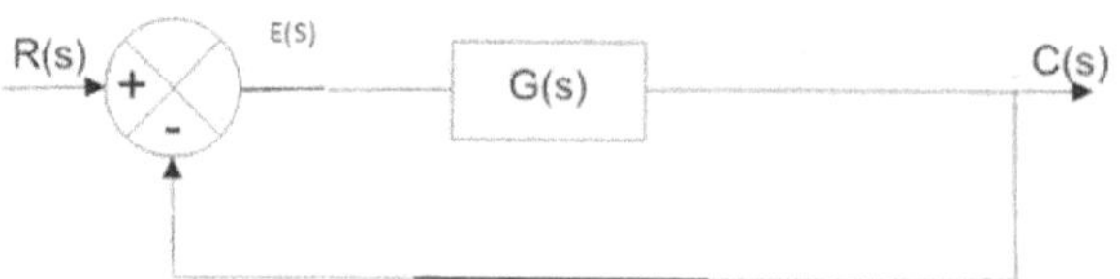

Fig.4.8 Block diagram representation of a second order control system.

The time response of a second order control system subjected to unit step input function is given by:

$$c(t) = 1 - \frac{e^{-\zeta\omega_n t}}{\sqrt{1-\zeta^2}} \sin\left[(\omega_n \sqrt{1-\zeta^2})t + \tan^{-1}(\frac{\sqrt{1-\zeta^2}}{\zeta}) \right]$$

The error is given as c(t) = r(t)-c(t)

and therefore

$$e(t) = 1 - \frac{e^{-\zeta\omega_n t}}{\sqrt{1-\zeta^2}} \sin\left[(\omega_n \sqrt{1-\zeta^2})t + \tan^{-1}(\frac{\sqrt{1-\zeta^2}}{\zeta}) \right] \begin{pmatrix} a_1 & & 0 \\ & \ddots & \\ 0 & & a_n \end{pmatrix}$$

The steady state error is

$$e_{ss} = \lim_{t \to \infty} - \frac{e^{-\zeta\omega_n t}}{\sqrt{1-\zeta^2}} \sin\left[(\omega_n \sqrt{1-\zeta^2})t + \tan^{-1}(\frac{\sqrt{1-\zeta^2}}{\zeta}) \right]$$

The time response expression i.e., c(t) indicates that for value of $\varsigma < 1$the response presents exponentially decaying oscillations having a frequency ω_d and the time constant of exponential decay is $\frac{1}{\varsigma\omega_n}$. From the above equations the time response indicates that for values of $\varsigma < 1$ presents damped oscillations and such a response is called under damped response. If $\varsigma = 0$ the time response in relation to equation the output gives sustained oscillation are also called undamped oscillations. If $\varsigma = 1$ the time response is said to be critically damped and if $\varsigma > 1$ the time response is said to be over damped that produces sluggish response. However based on the characteristic equation of second order system the roots are considering as a poles of transfer function. Therefore the study of roots gives a prediction about the nature of time response. The real part of the roots represents the damping and imaginary part represents damped frequency of oscillations. So the location of roots of characteristic equation for various values of ς keeping ω_n fixed and the corresponding time response of second order control system as shown in Fig.4.9.

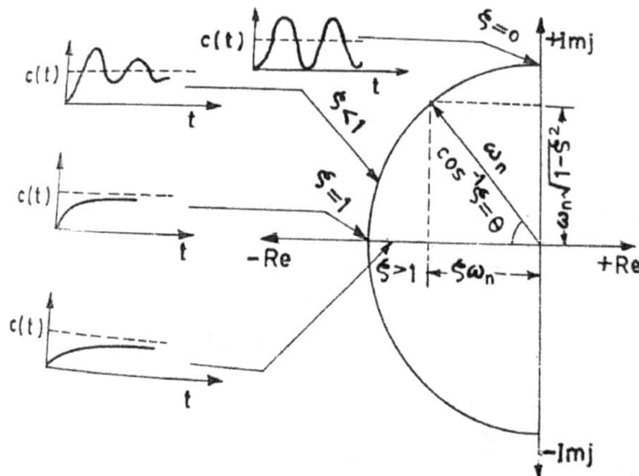

Fig.4.9 Location of roots of the characteristic equation and corresponding time response.

1. ***Time domain specifications***: The time response of an under damped control system exhibits damped oscillations prior to reaching steady state. The specifications pertaining to time response during transient part are expressed in terms of definition and mathematical expressions are described below.

1. **Delay Time (t_d):** It is the time required for the response to reach 50% of the final value for the final value for the first time. It is expressed as,

$$t_d = \frac{1+0.7\varsigma}{\omega_n}$$

2. **Time constant (τ):** Time constant is the time taken for the system to reach 63.2% of its final value for the first time. It is evident that the magnitude of oscillation decreases as $e^{-\varsigma\omega_n t}$. According to the definition, the output reaches 63.2% of its final value when $t = \frac{1}{\varsigma\omega_n}$.

   Hence the time constant of a second order system is $T = \frac{1}{\varsigma\omega_n}$

3. **Rise time (t_r):** Rise time is defined as the time taken for the response to rise from 0 to 100% of its final value. In equation putting $t=t_r$ and c(t)=1 and R=1, we get the following equation.

$$C(t_r)=1=1[-[-\frac{e^{-\varsigma w_n t_r}}{\sqrt{1-\zeta^2}}\sin(w_d t_r+\phi)]$$

$$t_r = \frac{(\pi - \phi)}{\omega_n\sqrt{1-\varsigma^2}} \quad \text{where } \varphi \text{ is in radians}$$

4. **Overshoot (M_p):** The overshoot of a system indicates the relative stability of the system. It is often expressed as percentage or percentage overshoot.

   Peak overshoot M_p = (Maximum peak value – Final value)

$$\% \text{ Peak overshoot } = \% M_p = \frac{(\text{Maximum peak value} - \text{Final value})}{\text{Final Value}} \times 100$$

5. **Settling time (t_s):** It is the time required for the response to decreased to and stay within a specified percentage of final value.

$$t_s = \frac{m}{\varsigma\omega_n} = mT$$

6. **Steady-state error constants:** As the steady state error is an index of accuracy of a control system. Therefore the steady state error should be minimum as far as possible. The steady state performance of a control system is assessed by the magnitude of the steady state error expressed by the system and the system input specified as either step or ramp or parabolic.

The magnitude of the steady state error in a closed loop control system depends on its open loop transfer function i.e., G(s).H(s) of the system. Therefore the steady state error can be determined as

$$e(\infty) = \lim_{s\to 0} sE(s) = \lim_{s\to\infty} \frac{sR(s)}{1+G(s)}$$

From the above equation the steady state error possessed by a closed loop control system depends on the input and the open loop transfer function. The actual output of a control system may be in any physical form is called as 'position' or displacement. The first derivative of the actual output is called 'velocity' and the second derivative is 'acceleration'.

For evaluating steady state error the input function is specified as either unit step(displacement) or unit ramp(velocity) or unit parabolic(acceleration) and it is also depends on the type of open loop transfer function of a closed loop control system. So the steady state error as estimated based on the type of an open loop transfer function with the specified input is summarized in Table 4.2.

Table 4.2 Static error co-efficient for different type of systems and input

Input	$\mathcal{L}$(input)	Type '0'		Type '1'		Type '2'	
		Static error coeff	e_{ss}	Static error coeff	e_{ss}	Static error coeff	e_{ss}
Unit step	$\frac{1}{s}$	K_p=K	$\frac{1}{1+K}$	$K_p=\infty$	0	$K_p=\infty$	0
Unit ramp	$\frac{1}{s^2}$	K_v=0	∞	K_v=K	$\frac{1}{K}$	$K_v=\infty$	0
Unit parabolic	$\frac{1}{s^3}$	K_a=0	∞	K_a=0	∞	K_a=K	$\frac{1}{K}$

An observation of the above table gives the following information:

(i) For type '0'system the ramp and parabolic inputs are not acceptable.
(ii) For type '1' system the parabolic input are not acceptable.
(iii) For type '2' system all the three inputs the step, ramp and parabolic inputs are acceptable.
(iv) Finite steady state error varies in inverse proportion to the forward path gain k.
(v) All the steady state errors are positional in nature.

EXPERIMENTS

4.10 Time Response of Second Order Control System

Aim: To determine time domain specifications of the given transfer function

Apparatus: Personal Computer (PC)

MATLAB software

THEORY:

The time response has utmost importance for the design and analysis of control systems because these are inherently time domain systems where time is independent variable. During the analysis of response, the variation of output with respect to time can be studied and it is known as time response. To obtain satisfactory performance of the system with respect to time must be within the specified limits. From time response analysis and corresponding results, the stability of system, accuracy of system and complete evaluation can be studied easily.

Due to the application of an excitation to a system, the response of the system is known as time response and it is a function of time. The two parts of response of any system:

(i) Transient response

(ii) Steady-state response.

Transient response: The part of the time response which goes to zero after large interval of time is known as transient response.

Steady state response: The part of response that means after the transients have died out is said to be steady state response.

The total response of a system is sum of transient response and steady state response:

C(t) = Ctr(t) + Css(t)

Where C(t) = total response

Ctr(t) = transient response

Css(t) = steady state response

4.11 Time Response of Second Order Control System

A second order control system is one wherein the highest power of 's' in the denominator of its transfer function equals 2.

Transfer function is given by:

$$T(s) = \frac{\omega_n^2}{s^2 + 2\xi w_n s + \omega_n^2}$$

where, ω_n is called natural frequency of oscillations.

$\omega_d = \omega_n \sqrt{1-\delta^2}$ is called damping frequency oscillations.

δ affects damping and called damping ratio.

The time response of second order system is as shown below.

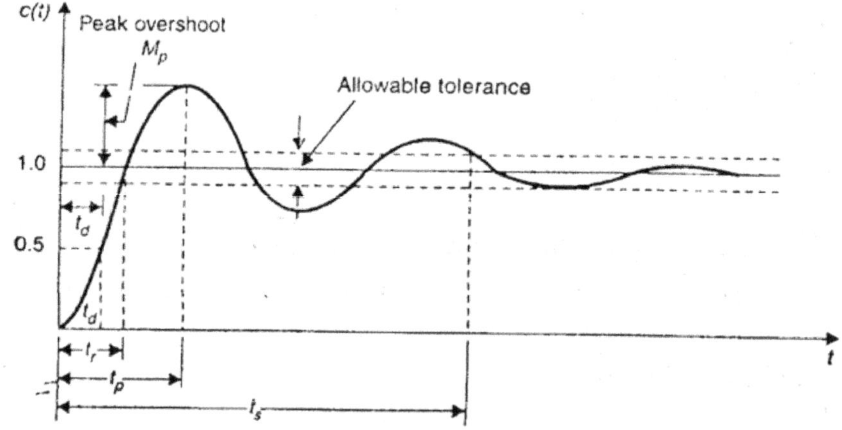

The expressions of different time domain specifications are given below.

Peak time $(t_P) = \dfrac{\pi}{\omega_d}$

Rise time $(t_r) = \dfrac{\pi - \tan^{-1}\sqrt{\dfrac{1-\zeta^2}{\zeta}}}{\omega_d}$

Settling time $(t_d) = \dfrac{3}{\zeta\omega_n}$

Damped frequency $(\omega_d) = \omega_n\sqrt{1-\zeta^2}$

Delay time $(t_d) = \dfrac{1+0.7\xi}{W_n}$

Peak Over shoot $(M_p) = \dfrac{e^{-\pi\xi}}{\sqrt{1-\xi^2}}$

MATLAB Program

```
clc;
Close all;
Clear all;
ωn= input ('Enter the value of ωn = ')
```

```
Zeta = input('Enter the value of zeta= ')
ω = sqrt(1-zeta^2);
ωd = ωn * ω;
d = -(pi*zeta);
delay time td=(1+(0.7*zeta))/ ωn
peak time tp=pi/ ωd
S=pi-acos(zeta);
Rise time tr=S/ ωd
Maximum peak over shoot Mp = exp(d/W)
Settling time ts =4/(zeta* ωn)
P = tf(4, [1 4 16]);
t=0:0.01:2;
step(P)
```

Example: Determine time domain specifications of the given transfer function $T(s) = \frac{4}{S^2 + 4S + 16}$

Output: Delay time (t_d) = 0.3375 sec

Peak time (t_p) = 0.9080 sec

Rise time(t_r) = 0.6053 sec

Maximum peak overshoot(Mp) = 0.1630

Settling time (t_s) = 2 sec

Time Response

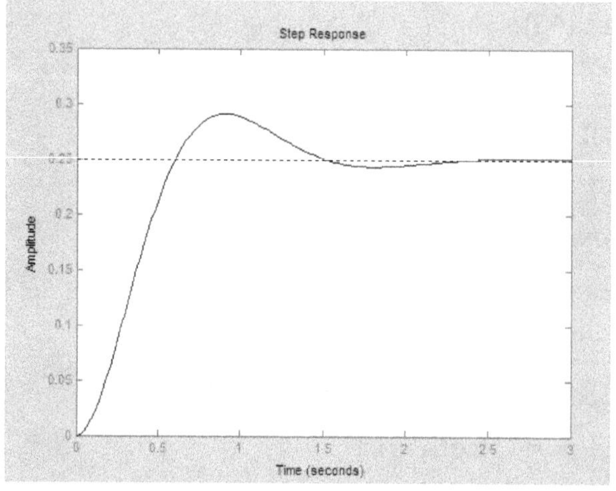

Result:

Viva-Voce:

1. What is transient response?
2. What is steady state response?
3. What is peak overshoot?
4. What is Peak time?
5. What is settling time?
6. What is rise time?
7. Explain the effect of damping ratio on the system response.

4.12 Step Response from a Given Transfer Function

Aim: To obtain the step response of a transfer function of the given system using MATLAB.

Apparatus: Personal Computer(PC)

MATLAB software

THEORY: A step signal is a signal whose value changes from one level to another level in zero time. Mathematically, the step signal is represented as given below:

R(t) = u(t) where

$$r(t) = \begin{Bmatrix} 0, & t & < & 0 \\ 1, & t & \geq & 0 \end{Bmatrix}$$

In the Laplace transform form,

$$R(s) = \frac{1}{S}$$

The step response of the given transfer function is obtained as follows:

$$T(s) = \frac{C(S)}{R(S)}$$

and $C(S) = T(S)*R(S)$

$$C(S) = = \frac{T(S)}{S}$$

Therefore the output is given by:

$$C(t) = L^{-1}[C(s)]$$

MATLAB Program:

```
clc
clear all
num = input('enter the numerator of the transfer
function')
den = input('enter the denominator of the transfer
function')
t=0:0.01:30;
c1=step(num,den,t);
plot(t,c1);
xlabel('Time in sec');
ylabel('c(t)');
title('Step response of open loop system');
```

Example:

Obtain the step response of the given transfer function

$$T(s) = \frac{4}{S^2 + 4S + 16}$$

Output:

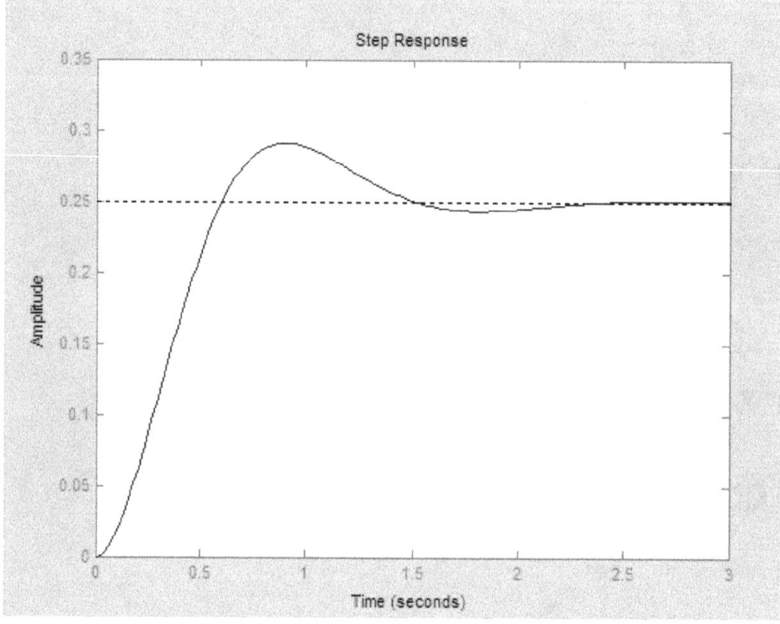

Result:

Viva-Voce:

1. What do you understand by MATLAB software? What does MATLAB stands for?
2. What is transfer function?
3. Define Step response.
4. Give the importance of step response.

4.13 Impulse Response from a given Transfer Function

Aim: To obtain the ramp response of a transfer function of the given system using MATLAB.

Apparatus: Personal Computer(PC)

MATLAB software

THEORY: An impulse signal is a signal whose value changes from zero to infinity in zero time. Mathematically, the unit impulse signal is represented as given below:

r(t) = δ(t) where

$$\begin{Bmatrix} 1, & t & = & 0 \\ 0, & t & \neq & 0 \end{Bmatrix}$$

In the Laplace transform form,

R(s) = 1

The step response of the given transfer function is obtained as follows:

$$T(s) = \frac{C(S)}{R(S)}$$

and C(S) = T(S)*R(S)

C(S) = 1*T(S)

Therefore the output is given by:

$$c(t) = L^{-1}[C(s)]$$

MATLAB Program:

```
clc
clear all
num = input('enter the numerator of the transfer function')
den = input('enter the denominator of the transfer function')
t=0:0.01:30;
c1=impulse(num,den,t);
plot(t,c1);
xlabel('Time in sec');
ylabel('c(t)');
title('impulse response of open loop system');
```

Example: Obtain the step response of the given transfer function

$$T(s) = \frac{4}{S^2 + 4S + 16}$$

Output:

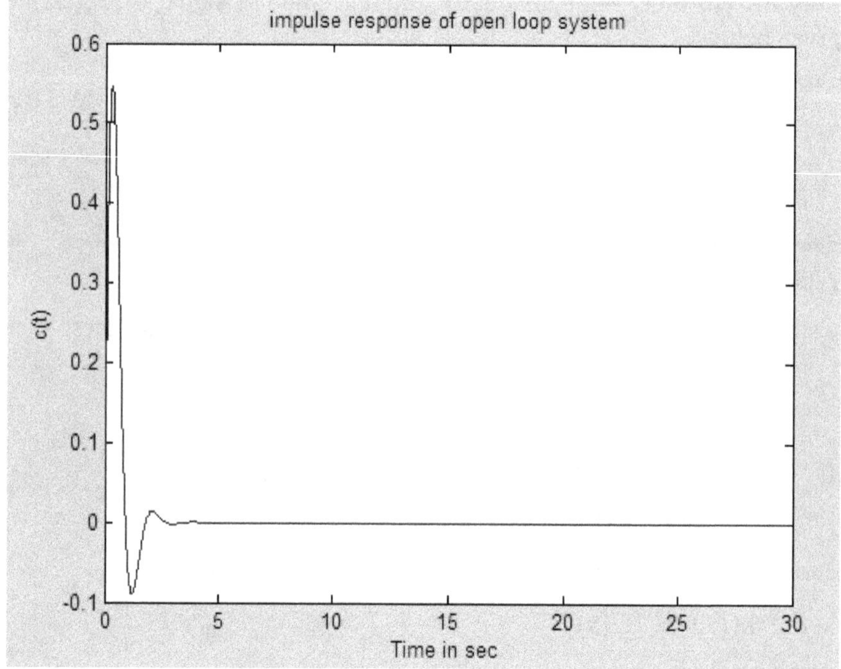

Result:

Viva-Voce:

1. Define Impulse response
2. Mention the test signals used to determine time response.
3. Define type and order of a system

4.14 Steady State Error Constants

Aim: To determine the Steady state error constants for different inputs

Apparatus Required: Personal Computer (PC)

MATLAB software

THEORY: Steady-state error is defined as the difference between the input (command) and the output of a system in the limit as time goes to infinity (i.e., when the response has reached steady state). The steady-state error will depend on the type of input (step, ramp, etc.) as well as the system type (0, I, or II).

We can calculate the steady-state error for this system from either the open- or closed-loop transfer function using the Final Value Theorem such as:

$$e(\infty) = \lim_{s \to 0} sE(s) = \lim_{s \to \infty} \frac{sR(s)}{1+G(s)}$$

$$e(\infty) = \lim_{s \to 0} sE(s) = \lim_{s \to o} sR(s)\left[1 - T(s)\right]$$

Now, let's plug in the Laplace transforms for some standard inputs and determine equations to calculate steady-state error from the open-loop transfer function in each case.

1. Step Input ($R(s) = 1 / s$):

$$e(\infty) = \frac{1}{1 + \lim_{s \to 0} G(s)} = \frac{1}{1 + Kp}$$

   Where $Kp = \lim_{s \to 0} G(s)$.

2. Ramp Input ($R(s) = 1 / s^2$):

$$e(\infty) = \frac{1}{\lim_{s \to 0} sG(s)} = \frac{1}{K_a}$$

   where $K_a = \lim_{s \to 0} sG(s)$.

3. Parabolic Input ($R(s) = 1 / s$^3):

$$e(\infty) = \frac{1}{\lim_{s \to 0} s^2 G(s)} = \frac{1}{K_a}$$

   where $K_a = \lim_{s \to 0} s^2 G(s)$

MATLAB Program:

```
clc
clear all;
```

```
num=input('enter the numerator of the transfer function')
den=input('enter the denominator of the transfer function')
g=tf(num,den)
t=feedback(1,[g])
disp('step input');
kp=dcgain(g)
ess=1/(1+kp)
disp('ramp input');
nsg=conv([1 0],n);
dsg=poly([1 1 0]);
sg=tf(nsg,dsg)
sg=minreal(sg)
kv=dcgain(sg)
ess=1/kv
disp('parabolic input');
ns2g=conv([1 0 0],n);
ds2g=poly([1 1 0]);
s2g=tf(ns2g,ds2g)
s2g=minreal(s2g)
ka=dcgain(s2g)
ess=1/ka
step(t);
```

Example: Determine steady state error constants of unity feedback system whose open loop transfer function $G(s) = \frac{1}{s(s+1)}$

Output:

Transfer function: $\frac{1}{s^2+s}$

Closed Loop Transfer function: $\frac{s^2+s}{s^2+s+1}$

Step input: kp = Inf

ess = 0

Ramp input:

Transfer function: $\frac{s}{s^3-2s^2+s}$

Transfer function: $\frac{1}{s^2-2s+1}$

kv = 1

ess = 1

Parabolic input:

Transfer function: $\frac{s^2}{s^3-2s^2+s}$

Transfer function: $\frac{s}{s^2-2s+1}$

ka = 0

ess = Inf

Result

Viva-Voce

1. Which type of input positional error constant K_p is indicative in output.
2. Which type of input velocity error constant K_v is inductive in output.
3. What is steady state error?
4. Define acceleration error constant.
5. What is the advantage of dynamic error constants.

Exercise Problems

1. For the following closed loop transfer functions obtain the Step/impulse/ramp responses of the given system and also obtain the Time domain specifications

   where $\zeta = 0.6,\ w_n = 8$

   (a) $$\frac{C(s)}{R(s)} = \frac{\omega_n^2}{S^2 + 2\zeta\omega_n S + \omega_n^2}$$

   (b) $$\frac{C(s)}{R(s)} = \frac{10s + 25}{(s+1)s^2 + 6S + 100}$$

2. Determine the Step, ramp and parabolic error constants and the steady state error for the feedback control systems with the following open loop transfer functions also plot the respective time domain responses.

   (a) $$G(s) = \frac{5S^2 + 2S + 1}{S(S+2)(S+8)(S+45)}$$

   (b) $$G(s) = \frac{150}{S^2 + 30S + 150}$$

5 Stability Analysis

5.1 Introduction

At first instance stability analysis requires a qualitative assessment that is if a system characteristic equation or transfer function of the system is given then without obtaining a direct time solution of characteristic equation. How one can say whether the system is stable or unstable in other words to get a YES or NO answer pertaining to stability means qualitative analysis the most important requirement in the design of feedback control system is system stability. Instability in the system produces undesirable effects and affects the system performance. Hence, stability is the fundamental property of the system and related studies constitute major part of the course content in control systems. A control system is tested initially to determine its absolute stability. A control system is said to be in the stable state when the output stays in the same state when no disturbance or input is applied to the system. If the system is found stable then it is necessary to know the stability strength or degree of stability which in other words is called relative stability. From engineering point of view, when we represent a system in block diagram, it is desirable to introduce the concept of input to the system gives bounded output. Otherwise the system is unstable.

5.2 Definitions of Stability

An adequate definition of stability with reference to linear control systems is given below.

'If any oscillation set up in a system is a consequence to application of an input are damped out with respect to time. The system is said to be stable.'

Conversely for unstable systems oscillations are increasing in magnitude if the magnitudes of the oscillations are sustained the system is marginally stable.

5.3 Absolute and Relative Stability

System is normally classified as absolutely stable and relatively stable. The answer for absolute stability is either yes or no. On the other hand, if tries to establish how stable it is, then degree of stability is the measure of relative stability. In the time domain, relative stability is measured by parameters such as maximum over shoot and damping ratio. Resonant peak is measure of relative stability in frequency domain. Higher the maximum overshoot and resonant peak lower the relative stability. Relative stability i the frequency domain can be measured using Nyquist plot. For a minimum phase T.F. if the Nyquist plot is nearer to (–1, 0) point, the relative stability of the system is less. Gain margin and phase margin are also a measure of relative stability in the frequency domain. Higher the phase margin and gain margin, higher is the relative stability.

5.4 Zero Input Stability

Consider a system which is driven by its initial conditions only, and there is no input applied. Under such conditions, if the output response c(t) of the system reaches zero as time t approaches infinity, the system is said to be zero input stable.

5.5 Asymptotic Stability

When the system is driven by its initial condition with zero input applied and if the output response c(t) of the system reaches zero as time t approaches infinity, the system is said to be asymptotically stable. Zero input stability is also called asymptotic stability.

5.6 Marginal Stability

For BIBO, zero input asymptotic stability, in all these cases, the general requirement is that the roots of the characteristics equation fall in the left half of the s-plane. If some roots of the characteristics equation fall in imaginary (jω) axis, while all the other roots lie in LHP, the system is said to be marginally stable. It is to be noted that the system with root located at the origin (pure integrator with s=0) is consider as a stable system. In the case of oscillators, the roots are located on the (jω) axis and here also the system is considered to be stable. In control systems the following techniques are used to analyse stability of the system.

1. R-H Criteria
2. Root Locus
3. Polar plot
4. Nyquist Criteria

5. Bode Plot
6. M and N circles
7. Nichols charts

The first two are analytical methods where the remaining all are graphical methods are also called as frequency plots. Moreover in control system experimental point of view most useful methods are R-H Criteria, Root Locus, Nyquist Criteria and Bode Plot which are described below.

5.7 R-H Criteria

While applying R-H criteria it is imperative that no powers of 's' in the characteristic equation be absent, any absence of such a power indicates the presence of alter one positive real part root and confirms system instability by inspection. However, If the characteristic equation contains either only odd or even powers os 's'. This indicates that the roots have no real parts and possess only imaginary parts and therefore the system has sustained oscillations in its output response. The following difficulties are faced while applying R-H criteria.

(i) The element in one of the columns of routh array is zero

Remedy: This can be overcome by substituting small □ value in place of zero or by replacing '1/z' inplace 's' in the given characteristic equation.

(ii) All the elements in one of the rows of routh array are zero.

Remedy: It can be overcome by taking upper row elements of routh array as a auxillary polynomial expression by differentiating with respect to s and equate it to zero.

5.8 Root Locus

The stability of a closed loop control system is determined from the location of the roots b of the characteristic equation. For a system to be stable the roots of this characteristic equation should be located in the L.H.S of s-plane.

However, closed loop control system having forward path gain K as shown in block diagram Fig.5.1.

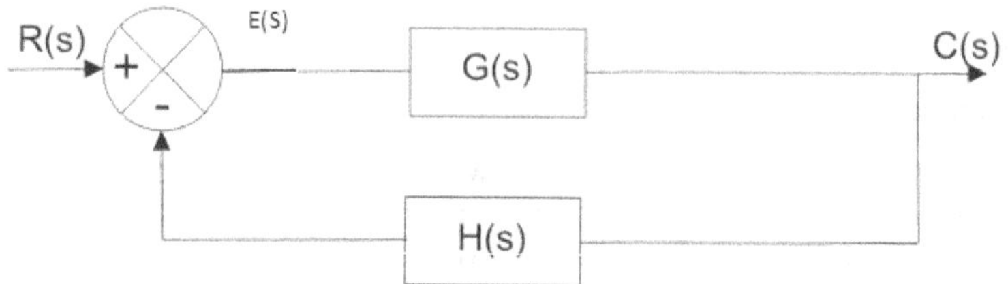

Fig.5.1 Block diagram of closed loop control system.

The characteristic equation is given by 1 + G(s) H(s) = 0.

Any root of this equation satisfies the following two conditions.

$$|G(s)H(s)| = 1$$

and

$$\angle G(s)H(s)=(2k+1)180^0$$

where k = 0, 1, 2….

The root locus method of analysis is a process of determining the points in s-plane satisfying the above two conditions. Usually the forward path gain factor K is consider as in independent variable and the roots of 1 + G(s)H(s) = 0 as dependent variables, the roots are plotted in s-plane with K as a variable parameter. From the location of the roots in s-plane the nature of time response and system stability can be ascertained.

5.9 Nyquist Criteria

Nyquist Criteria is used to identify the presence of roots of a characteristic equation of a control system in a specified medium of s-plane. From the stability view point the specified region being the entire R.H.S beyond the imaginary axis of complex s-plane. Although the purpose of using Nyquist Criteria is similar to R-H criteria but the approach differs in following respects:

(i) The open loop transfer function G(s)H(s) is consider instead of closed loop charactestic equation 1+ G(s)H(s)=0.

(ii) Inspection of graphical plot of G(s)H(s) enables to get more than YES or No answer of R-H method pertaining to stability of control systems.

If the Nyquist path in the s plane encircles Z zeros and P poles of 1 + G(s) H(s) and does not pass through any poles or zeros of 1 + G(s) H(s) as a representative point 's' moves in the clockwise direction along the Nyquist path, then the corresponding contour in the G(s)H(s) plane encircles the $-1 + j0$ point $N = Z - P$ times in the clock-wise direction i.e., Negative values of N imply counter clockwise encirclements.

In examining the stability of linear control systems using the Nyquist stability criterion, we see that three possibilities can occur:

1. There is no encirclement of the $-1 + j0$ point. This implies that the system is stable, if there are no poles of G(s)H(s) in the right-half 's' plane; otherwise, the system is unstable.
2. There are one or more counter clockwise encirclements of the $-1 + j0$ point. In this case the system is stable, if the number of counter clockwise encirclements is the same as the number of poles of G(s)H(s) in the right-half s plane; otherwise, the system is unstable.
3. There are one or more clockwise encirclements of the $-1 + j0$ point. In this case the system is unstable.

5.10 Bode Plot

The bode plot method gives a graphical procedure for determining the stability of a control system based on sinusoidal frequency response. The transfer function of a system for sinusoidal input response can be obtaining by substituting jω in place of laplace operator 's'. Therefore, If the open loop transfer function of a system G(s)H(s) the corresponding sinusoidal open loop transfer function G(jω)H(jω) which can be expressed in the form of magnitude and phase angle.

A Bode diagram consists of two graphs: One is a plot of the logarithm of the magnitude of a sinusoidal transfer function; the other is a plot of the phase angle; both are plotted against the frequency on a logarithmic scale.

The main advantage of using the Bode diagram is that multiplication of magnitudes can be converted into addition. Furthermore, a simple method for sketching an approximate log-magnitude curve is available. It is based on asymptotic approximations. Such approximation by straight-line asymptotes is sufficient if only rough in-formation on the frequency-response characteristics is needed. Should the exact curve be desired, corrections can be made easily to these basic asymptotic plots. Expanding the low-frequency range by use of a logarithmic scale for the frequency is highly advantageous, since characteristics at low frequencies are most important in practical systems. Although it is not possible to plot the curves right down to zero frequency because of the logarithmic frequency ($\log 0 = \infty$), this does not create a serious problem.

The relative stability of a closed loop control system can be conveniently assessed by plotting its open loop transfer function by bode plot methods.

The Gain Margin (GM) and Phase Margin (PM) can be determined directly from the bode plot. If both GM and PM are positive then the system is said to be stable. If any one of the GM and PM are negative then the system is said to be stable.

EXPERIMENTS

5.11 Root Locus from a Transfer Function

Aim: To plot the root locus for a given transfer function of the system using MATLAB.

Apparatus: Personal Computer (PC)

MATLAB software

Theory: As we change gain, we notice that the system poles and zeros actually move around in the S-Plane. This fact can make life particularly difficult, when we need to solve higher-order equations repeatedly, for each new gain value. The solution to this problem is a technique known as **Root-Locus** graphs. Root-Locus allows you to graph the locations of the poles and zeros *for every value of gain.* Root locus technique in control system is easy to implement as compared to other methods. With the help of root locus one can easily predict the performance of the whole system. Root locus provides the better way to indicate the parameters.

rlocus computes the Evans root locus of a SISO open-loop model. The root locus gives the closed-loop pole trajectories as a function of the feedback gain k (assuming negative feedback). Root loci are used to study the effects of varying feedback gains on closed-loop pole locations. In turn, these locations provide indirect information on the time and frequency responses.

rlocus(sys) calculates and plots the root locus of the open-loop SISO model sys. This function can be applied to any of the following feedback loops by setting sys appropriately.

MATLAB Program:

```
Clc;

Clear all;

num=input('enter the numerator of the transfer function')
```

```
den=input('enter the denominator of the transfer function')
h=tf(num,den)
rlocus(h)
```

Example: Find root locus of the given transfer function

$$T(s) = \frac{S^3 + 5S^2 + 4S + 6}{4S^3 + 7S^2 + 12S + 9}$$

Output:

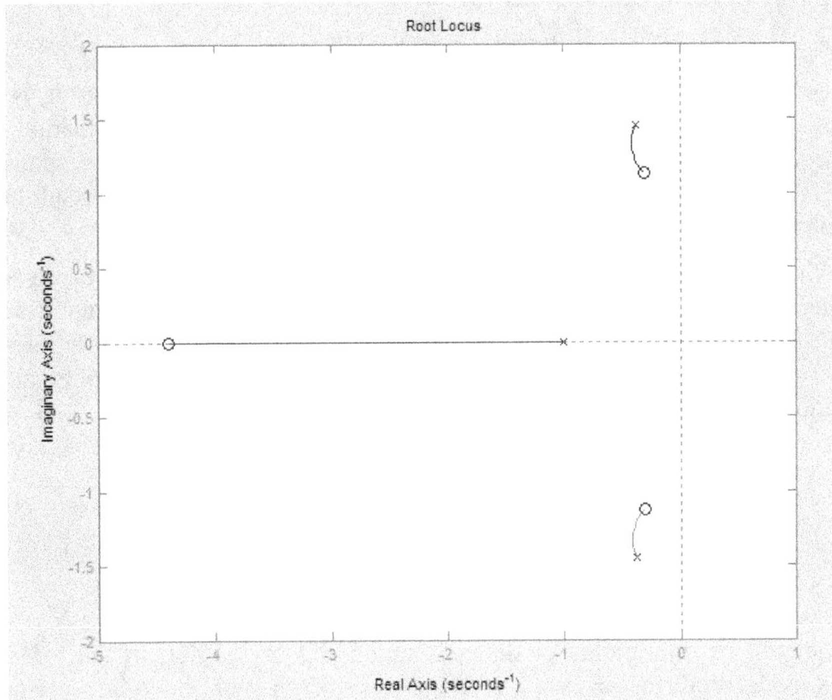

Result:

Viva-Voce:

1. What is the effect of root locus adding poles on a system?
2. What is the effect of adding zeros on a system?

3. What is the importance of root locus on stability?
4. What are the drawbacks of root locus?
5. To make system stable, the poles and zeros are in which way of s-plane.

5.12 Bode Plot From a Transfer Function

Aim: To obtain bode plot for a givan transfer function of the system using MATLAB.

Apparatus: Personal Computer (PC)

MATLAB software

Theory: In electrical engineering and control theory, a Bode plot is a graph of the frequency response of a system. It is usually a combination of a Bode magnitude plot, expressing the magnitude of the frequency response, and a Bode phase plot, expressing the phase shift. Both quantities are plotted against a horizontal axis proportional to the logarithm of frequency.

The magnitude axis of the Bode plot is usually expressed as decibels of power, that is by the 20 log rule: 20 times the common (base 10) logarithm of the amplitude gain. With the magnitude gain being logarithmic, Bode plots make multiplication of magnitudes a simple matter of adding distances on the graph (in decibels), since

$$\log (a \, . \, b) = \log (a) + \log (b).$$

A **Bode phase plot** is a graph of phase versus frequency, also plotted on a log-frequency axis, usually used in conjunction with the magnitude plot, to evaluate how much a signal will be Phase- shifted.

bode(sys) plots the Bode response of an arbitrary LTI model sys. This model can be continuous or discrete, and SISO or MIMO. In the MIMO case, bode produces an array of Bode plots, each plot showing the Bode response of one particular I/O channel. The frequency range is determined automatically based on the system poles and zeros.

bode(sys,w) explicitly specifies the frequency range or frequency points to be used for the plot. To focus on a particular frequency interval $[W_{min}, W_{max}]$, set w = $\{W_{min}, W_{max}\}$. To use particular frequency points, set w to the vector of desired frequencies. Use logspace to generate logarithmically spaced frequency vectors. All frequencies should be specified in radians/sec.

bode(sys1,sys2,...,sysN) or bode(sys1,sys2,...,sysN,w) plots the Bode responses of several LTI models on a single figure. All systems must have the same number of inputs and outputs, but may otherwise be a mix of continuous and discrete systems. This syntax is useful to compare the Bode responses of multiple systems.

bode(sys1,'PlotStyle1',...,sysN,'PlotStyleN') specifies which color, linestyle, and/or marker should be used to plot each system. For example, bode(sys1,'r--',sys2,'gx') uses red dashed lines for the first system sys1 and green 'x' markers for the second system sys2.

When invoked with left-side arguments

[mag,phase,w] = bode(sys)

[mag,phase] = bode(sys,w)

return the magnitude and phase (in degrees) of the frequency response at the frequencies w (in rad/sec). One can convert the magnitude to decibels by

magdb = 20*log10(mag)

MATLAB Program:

```
Clc;
Clear all;
num=input('enter the numerator of the transfer function')
den=input('enter the denominator of the transfer function')
h=tf(num,den)
[gm pm wcp wcg]=margin(h)
bode(h)
if(gm&pm>0)
disp('system is stable')
else
disp('system is unstable')
```

Example:

Find bode plot of the given transfer function and also find its gain and phase margin.

$$\frac{1.25}{0.5\,s^3 + 1.5\,s^2 + s}$$

Output:

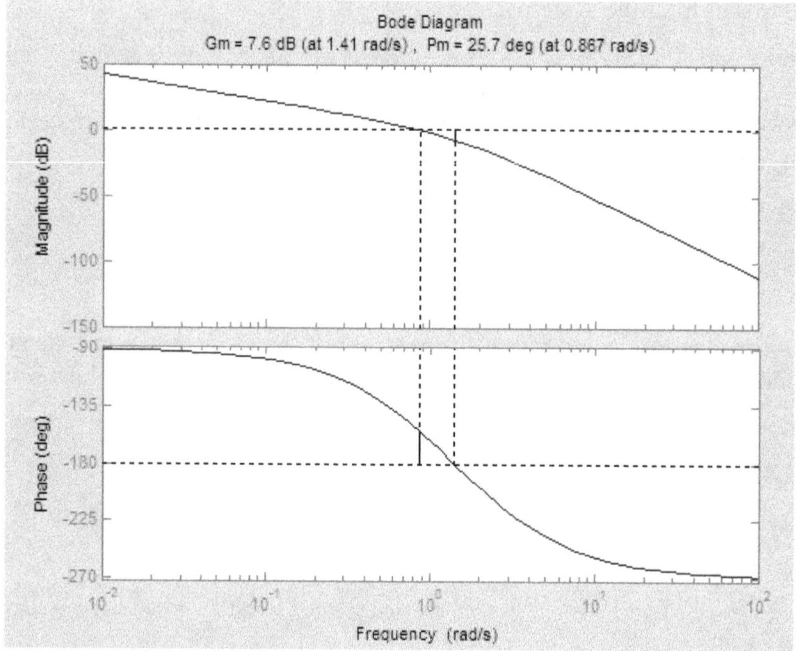


```
Gain Margin(GM)= 7.6 db.
Phase Margin (PM)= 25.7 deg.
System is stable.
```

Result:

Viva-Voce:

1. What is the gain margin of the 2nd order system?
2. What is phase cross over frequency?
3. What is gain cross over frequency?
4. How can you define stability of the system based on GM and PM.
5. How to find the Transfer function from the given bode plot.

5.13 Nyquist Plot From a Transfer Function

Aim: To obtain the Nyquist plot for a given transfer function of the system.

Apparatus: Personal Computer (PC)

MATLAB software

Theory: A nyquist plot is used in automatic control and signal processing for assessing the stability of a system with feedback. It is represented by a graph in polar coordinates in which the gain and phase of a frequency response are plotted. The plot of these phasor quantities shows the phase as the angle and the magnitude as the distance from the origin. This plot combines the two types of Bode plot magnitude and phase on a single graph with frequency as a parameter along the curve.

Nyquist calculates the Nyquist frequency response of LTI models. When invoked without left hand arguments, nyquist produces a Nyquist plot on the screen. Nyquist plots are used to analyze system properties including gain margin, phase margin, and stability.

The nyquist stability criterion, provides a simple test for stability of a closed-loop control system by examining the open-loop system's Nyquist plot. Stability of the closed-loop control system may be determined directly by computing the poles of the closed-loop transfer function. The Nyquist Criteria can tell us things about the frequency characteristics of the system. For instance, some systems with constant gain might be stable for low-frequency inputs, but become unstable for high-frequency inputs. Also, the Nyquist Criteria can tell us things about the phase of the input signals, the time-shift of the system, and other important information.

The Nyquist Contour: The nyquist contour, the contour that makes the entire nyquist critcrion work, must encircle the entire right half of the complex s plane. Remember that if a pole to the closed-loop transfer function (or equivalently a zero of the characteristic equation) lies in the right-half of the s plane, the system is an unstable system. To satisfy this requirement, the nyquist contour takes the shape of an infinite semi-circle that encircles the entire right-half of the s plane.

Nyquist Criteria: Let us first introduce the most important equation when dealing with the Nyquist criterion:

$$N = Z - P$$

where N is the number of encirclements of the $(-1, 0)$ point.

Z is the number of zeros of the characteristic equation.

P is the number of poles of the characteristic equation.

With this equation stated, one can now state the Nyquist Stability Criterion:

Nyquist Stability Criterion: A feedback control system is stable, if and only if the contour $\Gamma_{F(s)}$ in the F(s) plane does not encircle the (–1, 0) point when P is 0.

A feedback control system is stable, if and only if the contour $\Gamma_{F(s)}$ in the F(s) plane encircles the (–1, 0) point a number of times equal to the number of poles of F(s) enclosed by Γ. In other words, if P is zero then N must equal zero. Otherwise, N must equal P. Essentially, we are saying that Z must always equal zero, because Z is the number of zeros of the characteristic equation (and therefore the number of poles of the closed-loop transfer function) that are in the right-half of the s plane.

MATLAB Program:

```
Clc;
Clear all;
num=input('enter the numerator of the transfer function')
den=input('enter the denominator of the transfer function')
h=tf(num,den)
nyquist(h)
[gm pm wcp wcg]=margin(h)
if(wcp>wcg)
disp('system is stable')
else
disp('system is unstable')
end
```

Example: Find Nyquist plot of the given transfer function.

$$\frac{1.25}{0.5\,s^3 + 1.5 + s^2 + s}$$

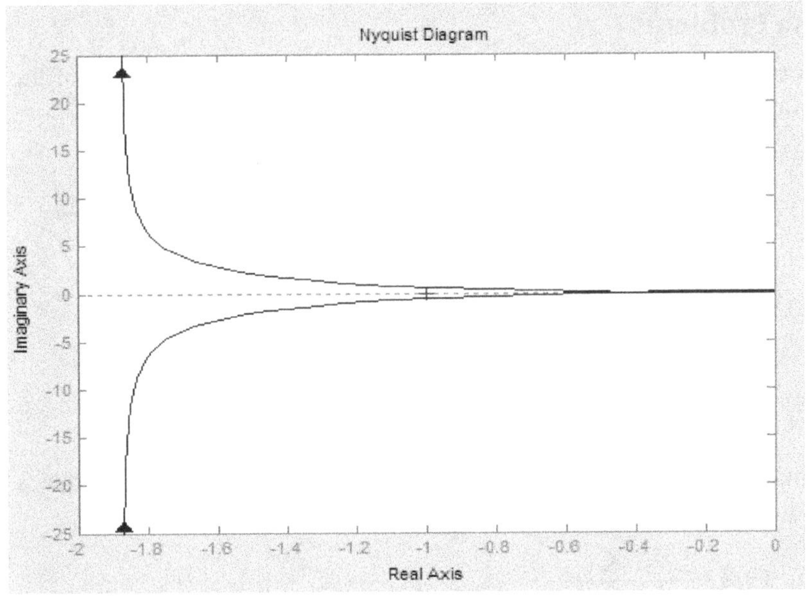

Output:

```
Gain Margin(GM) =     2.4000
Phase Margin(PM)=   25.6546
Phase Cross Over Frequency(wcp) =    1.4142
Gain Cross Over Frequency(Wcg) =    0.8667
system is stable
```

Result:

Viva-Voce:

1. Nyquist criterion is used to which type of stability? (absolute or relative or both)
2. What is the significance of anti-clock of encirclement in nyquist plot?
3. Can you use nyquist plot for non-linear control systems.
4. Significance of +ve and –ve encirclements.
5. By adding 'n' poles at origin, nyquist locus is rotated by an angle $\frac{n\pi}{2}$ in which direction at $w = 0$ and $w = \infty$ frequencies.

Exercise Problems

1. Obtain the Nyquist plot, Bode plot and determine the stability of the following unity feedback systems with open loop transfer functions are:

   (i) $G(s) = \dfrac{S+5}{S(S+1)(S+7)}$

   (ii) $G(s) = \dfrac{15(S+5)}{S(S+2)(S^2+6S+15)}$

   (iii) $G(s) = \dfrac{12(S+5)}{S^2-6S+10}$

2. Obtain the root locus plot for K > 0, the following unity feedback systems with open loop transfer functions are

   (i) $G(s) = \dfrac{K}{s(s+4)(s+8)}$

   (ii) $G(s) = \dfrac{K(s+7)}{(s+2)(s^2+2s+3)}$

   Also find the range of K for which the systems are stable. Also find the breakaway points.

6 Compensation and Controllers of Control Systems

Compensation and Controllers

PART-A

6.1 Introduction

If the performances of a control system is not up to the expectations as per desired specifications then it is required that some change in s system is needed to obtain the desired performance. The change can be in the form of adjustment of forward path gain or inserting a compensating device in control systems. The forward path gain adjustment is not suitable where in the system becomes unstable due to such adjustment. In such cases a compensation network is introduced in the system.

6.2 Necessary of Compensation

1. In order to obtain the desired performance of the system, we use compensating networks. Compensating networks are applied to the system in the form of feed forward path gain adjustment.
2. Compensate a unstable system to make it stable.
3. A compensating network is used to minimize overshoot.
4. These compensating networks increase the steady state accuracy of the system. An important point to be noted here is that the increase in the steady state accuracy brings instability to the system.
5. Compensating networks also introduces poles and zeros in the system thereby causes changes in the transfer function of the system. Due to this, performance specifications of the system change.

6.3 Types of Compensation Techniques

(a) ***Series compensation:*** Connecting compensating circuit between error detector and plants known as series compensation as shown in Fig.6.1.

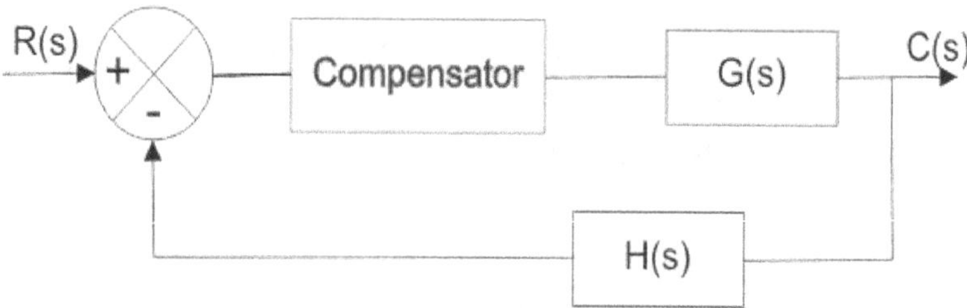

Fig.6.1 Series Compensator

(b) ***Feedback compensation***: When a compensator used in a feedback manner called feedback compensation as shown in Fig.6.2.

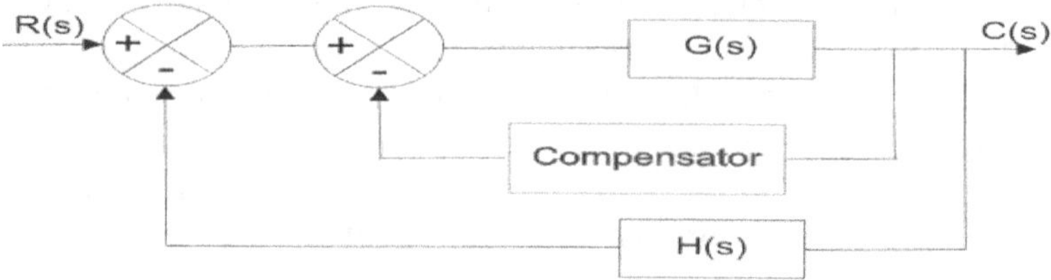

Fig.6.2 Feedback Compensator

(c) ***Load compensation***: A combination of series and feedback compensator is called load compensation as shown in Fig.6.3.

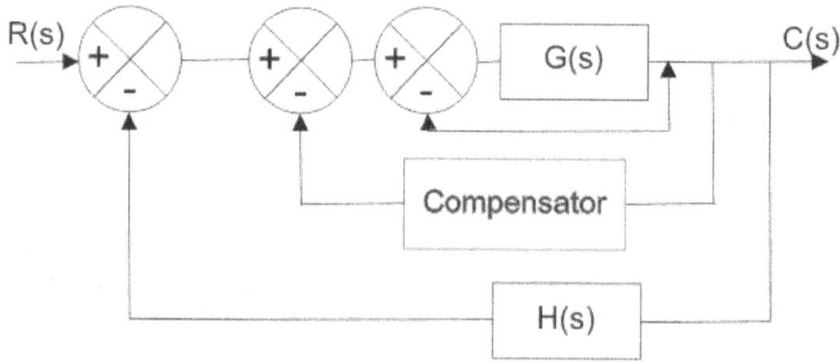

Fig.6.3 Load Compensator

Now what are compensating networks? A compensating network is one which makes some adjustments in order to make up for deficiencies in the system. Compensating devices are may be in the form of electrical, mechanical, hydraulic etc. Most electrical compensator are RC filter. The simplest network used for compensator are known as lead, lag network.

6.4 Phase Lead Compensation

A phase lead network where in the phase of output voltage leads the phase of input voltage for sinusoidal inputs. The transfer function of a lead network can be expressed in a sinusoidal form and the corresponding pole-zero configurations as shown in Fig.6.4 below and the experimental bode plot diagram discussed in experimental session through MATLAB program.

A system which has one pole and one dominating zero (the zero which is closer to the origin than all over zeros is known as dominating zero.) is known as lead network. If we want to add a dominating zero for compensation in control system then we have to select lead compensation network. The basic requirement of the phase lead network is that all poles and zeros of the transfer function of the network must lie on (-)ve real axis interlacing each other with a zero located at the origin of nearest origin.

$$G(s) = \frac{\alpha(1 + j\omega T)}{(1 + j\omega T)}$$

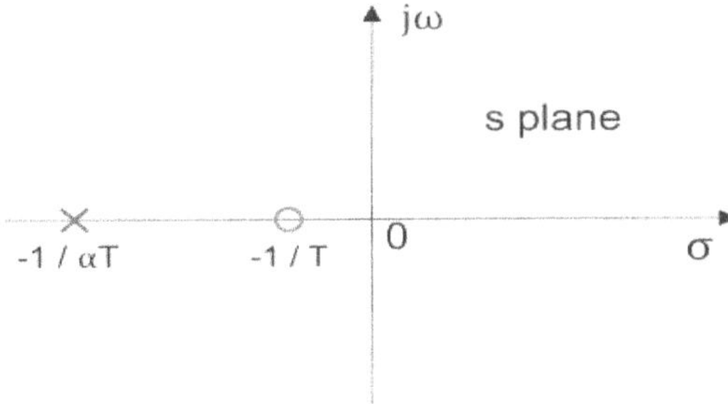

Fig.6.4 Pole-zero configuration of lead compensator

Effect of Phase Lead Compensation

1. The velocity constant K_v increases.
2. The slope of the magnitude plot reduces at the gain crossover frequency so that relative stability improves & error decrease due to error is directly proportional to the slope.
3. Phase margin increases.
4. Response become faster.

Advantages of Phase Lead Compensation

Let us discuss some of the advantages of the phase lead compensation-

1. Due to the presence of phase lead network the speed of the system increases because it shifts gain crossover frequency to a higher value.
2. Due to the presence of phase lead compensation maximum overshoot of the system decreases.

Disadvantages of Phase Lead Compensation

Some of the disadvantages of the phase lead compensation-

1. Steady state error is not improved.

6.5 Phase Lag Compensation

A phase lag network where in the phase of output voltage lags the phase of input voltage for sinusoidal inputs. The transfer function of a lag network can be expressed in a sinusoidal form and the corresponding pole-zero configurations as shown in Fig.6.5 below and the experimental bode plot diagram discussed in experimental session through MATLAB program.

A system which has one zero and one dominating pole (the pole which is closer to origin that all other poles is known as dominating pole) is known as lag network. If we want to add a dominating pole for compensation in control system then, we have to select a lag compensation network. The basic requirement of the phase lag network is that all poles & zeros of the transfer function of the network must lie in (-)ve real axis interlacing each other with a pole located or on the nearest to the origin.

$$G(s) = \frac{(1 + j\omega T)}{(1 + j\omega\, \beta T)}$$

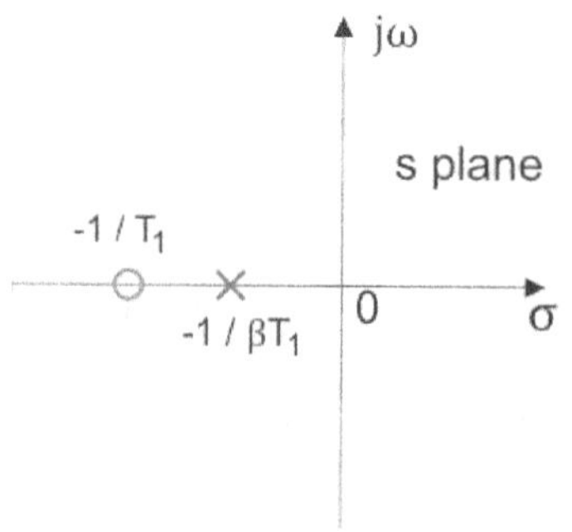

Fig.6.5 Pole-zero configuration of Lag compensator

Effect of Phase Lag Compensation

1. Gain crossover frequency increases.
2. Bandwidth decreases.
3. Phase margin will be increase.
4. Response will be slower before due to decreasing bandwidth, the rise time and the settling time become larger.

Advantages of Phase Lag Compensation

Let us discuss some of the advantages of phase lag compensation -

1. Phase lag network allows low frequencies and high frequencies are attenuated.
2. Due to the presence of phase lag compensation the steady state accuracy increases.

Disadvantages of Phase Lag Compensation

Some of the disadvantages of the phase lag compensation -

1. Due to the presence of phase lag compensation the speed of the system decreases.

6.6 Phase Lag-Lead Compensation

A single lag or lead compensation may not satisfy design specifications. For an unstable uncompensated system, lead compensation provides fast response but does not provide enough phase margins whereas lag compensation stabilize the system but does not provide enough bandwidth. For lead compensation shift the gain cross over frequency point to a higher value due to this the bandwidth is increased and improves the speed of response and overshoot is reduced. Similarly in lag compensation shift the gain cross over frequency point to a lower value due to this the bandwidth is decreased and improves the steady state error but the speed of response and overshoot is reduced. Therefore the speed response as well as steady state error of a control system can be simultaneously improved by both compensation networks are used. The transfer function of Lag-Lead compensator and corresponding pole-zero location is as shown in Fig.6.6 below.

$$G(s) = \frac{\alpha(1+j\omega T_1)}{(1+j\omega\alpha T_1)}\frac{(1+j\omega T_2)}{(1+j\omega\beta T_2)}$$

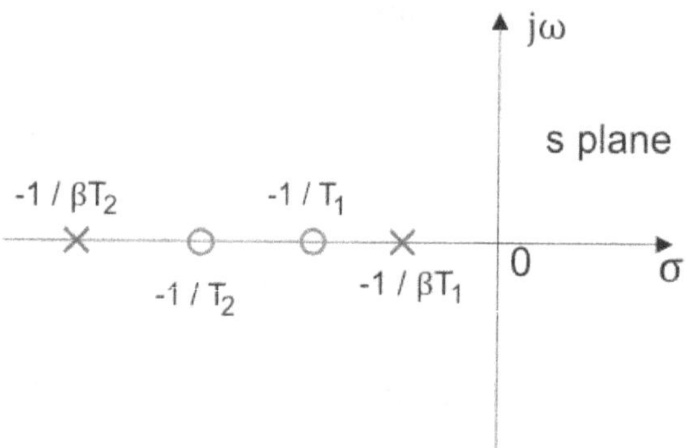

Fig.6.6 Pole-zero configuration of Lag-Lead compensator

Advantages of Phase Lag-Lead Compensation

Let us discuss some of the advantages of phase lag- lead compensation-

1. Due to the presence of phase lag-lead network the speed of the system increases because it shifts gain crossover frequency to a higher value.
2. Due to the presence of phase lag-lead network accuracy is improved.

PART-B

Controllers

6.7 Introduction

An automatic control system is used to maintain its output with in desirable limits by means of control action. Any deviation of the output from the reference input is detected by an error detector. The error thus detected is used as activating signal for control action through a controller. The proportional control action and types of control actions used for improving transient and steady state response of a control system are described below.

6.8 Types of Controllers

1. Proportional Controller (P- Controller)
2. Integral Controller (I- Controller)
3. Derivative Controller (D- Controller)
4. PI - Controller
5. PD - Controller

These controllers are considering as a PID (Proportional + Integral + Derivative) controller which produces an output signal consisting of three terms: one proportional to error signal, another one proportional to integral of error signal and third one proportional to derivative of error signal. The combination of proportional, Integral and derivative control action is called PID control action. A typical block diagram representation of the PID controller is as shown in Fig.6.7. The combined action has the advantage of each of the three individual control actions.

- The proportional controller stabilizes the gain the produces a steady state error.
- The integral controller reduces or eliminates the steady state error.
- The derivative controller reduces the rate of change of error.

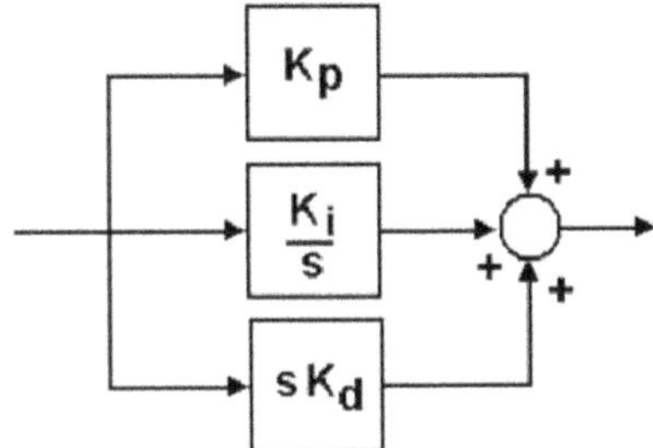

Fig.6.7 Typical Block diagram representation of PID controller.

The proportional, integral and derivative outputs are added together. The PID controller can be thought of as having a transfer function. The PID controller transfer function can be obtained by adding the three terms.

$$G_c(s) = K_p + K_i/s + sK_d$$

where $G_c(S)$ is called controller transfer function.

The transfer function can be combined into a pole-zero form.

$$G_c\,(s) = [sK_p + K_i + s^2K_d]/s$$

Since there is a quadratic in the numerator, there are two zeroes in this transfer function as well as the obvious pole at the origin, s = 0.

The PID controller transfer function really adds a pole at the origin, and two zeroes that can be anywhere in the s-plane that the designer wants, depending upon the designer's choice of the three gains. The block diagram of a control system with controller is as shown in Fig.6.8

$$G_c\,(s) = [sK_p + K_i + s^2K_d]/s$$

This gives the designer an incredibly large number of possibilities.

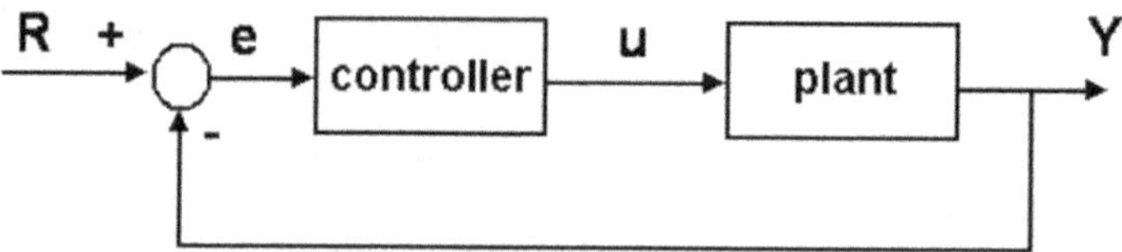

Fig.6.8 Block diagram of a control system with controller

Let's take a look at how the PID controller works in a closed-loop system using the schematic shown above. The variable (e) represents the tracking error, the difference between the desired input value (R) and the actual output (Y). This error signal (e) will be sent to the PID controller, and the controller computes both the derivative and the integral of this error signal. The signal (u) just past the controller is now equal to the proportional gain (K_p) times the magnitude of the error plus the integral gain (K_i) times the integral of the error plus the derivative gain (K_d) times the derivative of the error.

$$u = K_p e + K_r \int edt + K_D \frac{de}{dt}$$

This signal (u) will be sent to the plant, and the new output (Y) will be obtained. This new output (Y) will be sent back to the sensor again to find the new error signal (e). The controller takes this new error signal and computes its derivative and its integral again. This process goes on and on.

6.9 Characteristic Analysis of PID Controller

The PID controller characteristics are analyzed by considering proportional time constant (K_p), Integral Time constant (K_i) and derivative Time constant (K_d) of PID controller. A proportional controller (K_p) will have the effect of reducing the rise time and will reduce, but never eliminate, the steady-state error. An integral control (K_i) will have the effect of eliminating the steady-state error, but it may make the transient response worse. A derivative control (K_d) will have the effect of increasing the stability of the system, reducing the overshoot, and improving the transient response.

The Effects of each of controller time constants K_p, K_d, and K_i on a closed-loop system are summarized in the following Table 6.1.

Table 6.1 Effects of each of controller time constants K_p, K_d, and K_i on a closed-loop system

Time Constants	Rise Time	Overshoot	Settling Time	e_{ss}
K_p	Decrease	Increase	Small Change	Decrease
K_i	Decrease	Increase	Increase	Eliminate
K_d	Small Change	Decrease	Decrease	Small Change

The PID controller is very much similar to that of lag-lead network. However compare to the lag-lead network the PID controller has the superiority of having lag behaviour throughout the low frequency end and a lead behaviour throughout the high frequency end. Therefore from the filtering stand point the PD controller is a high pass filter, PI controller is a low pass filter and the PID controller is a band pass or band attenuate filter, depending on the values of the controller parameters.

EXPERIMENTS

6.10 Lag Compensator

Aim: To design a lag compensator for a closed loop system.

Apparatus: Personal Computer (PC)

MATLAB Software:

Theory: A lag compensator can be thought of in several different ways.

- First, a lag compensator is a device that provides phase lag in its' frequency response.
- If the compensator has phase lag - and never a phase lead - then there are implications about where the corner frequencies are in the Bode' plot.
- Other implications are that the phase lag compensator will have only certain types of pole-zero patterns in the s plane.

Lag compensators are sometimes the best controller to use to get a system to do what you want it to do. Like other compensators, lag compensators can be used to adjust frequency response by adding equal numbers of poles and zeroes to a systems. Those added singularities may possibly be manipulated to give better stability, better performance and general improvement.

We can manipulate TF, SS, and ZPK models using the arithmetic and model interconnection operations described in Operations on LTI Models and analyze them using the model analysis s, such as bode and step. FRD models can be manipulated and analyzed in much the same way you analyze the other model types, but analysis is restricted to frequency-domain methods.

Using a variety of design techniques, you can design compensators for systems specified with TF, ZPK, SS, and FRD models. These techniques include root locus analysis, pole placement, LQG optimal control, and frequency domain loop-shaping. For FRD models, you can either: Obtain an identified TF, SS, or ZPK model using system identification techniques. Use frequency-domain analysis techniques.

Other uses of FRD Models: FRD models are unique model types available in the Control System Toolbox collection of LTI model types, in that they don't have a parametric representation. In addition to the standard operations you may perform on FRD models, you can also use them to: Perform frequency domain analysis on systems with nonlinearities using describing functions. Validate identified models against experimental frequency response data.

MATLAB Program

```
clc;
clear all;
GM=input('Enter Required Gain margin');
PM=input('Enter Required Phase margin');
KV=input('Enter velocity error');
numg=input('Enter the numerator coefficients');
deng=input('Enter the denominator coefficients');
G=tf(numg,deng);
GM=input('Enter Required Gain margin');
PM=input('Enter Required Phase margin');
KV=input('Enter velocity error');
%TO FIND k,use sG(s)H(s)
K=KV/(dcgain(conv([1 0],numg),deng));
w=logspace(-1,2,100);
[mag,ang]=bode(K*numg,deng,w);
```

```
[gm,pm,wcg,wcp]=margin(mag,ang,w);
PH1=(PM+5)-pm;
PC=-180+PH1;
for k=1:length(ang);
   if ang(k)-PC<=0;
      M=mag(k);
      wf=w(k);
      break;
   end;
end;
wh=wf/10;
w1=wh/M;
kc=w1/wh;
lagc=tf(kc*[1 wh],[1 w1]);
% compensated system
CGCS=K*G*lagc;
S=tf([1 0],[0 1]);
KVF=dcgain(S*CGCS);
CLS=feedback(CGCS,1);
margin(CGCS)
```

Example: Given open loop transfer function $G(s) = \dfrac{1}{s(s+1)(0.5s+1)}$ it is designed to compensate the system so that static velocity error constant kv = 5sec –1. The phase margin is at least 40 and gain margin is atleast 10 db

Output: Transfer function of Uncompensated System

$$\frac{G = 1}{0.5s^3 + 1.5s^2 + s}$$

Continuous-time transfer function.

Bode plot for Uncompensated System:

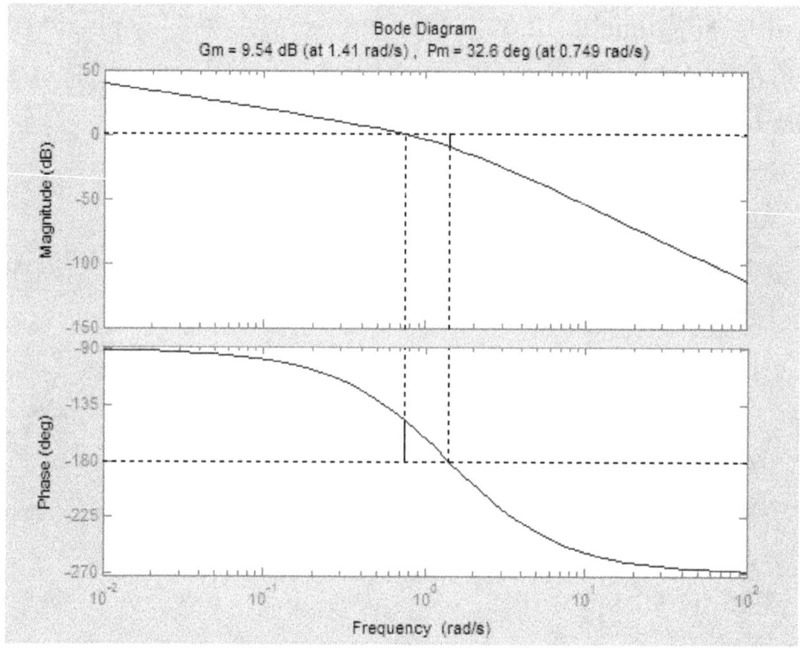

Gain Margin (GM) = 9.54 db

Phase Margin (PM) = 32.6 deg

Phase Cross Over Frequency(wcp) = 1.41 rad/sec

Gain Cross Over Frequency(Wcg) = 0.749 rad/sec

system is stable

Transfer function of compensated System

$$CGCS = \frac{0.4441s + 0.01793}{0.5\ s^4 + 1.502\ s^3 + 1.005\ s^2 + 0.003586\ s}$$

Continuous-time transfer function.

Bode plot for Compensated System:

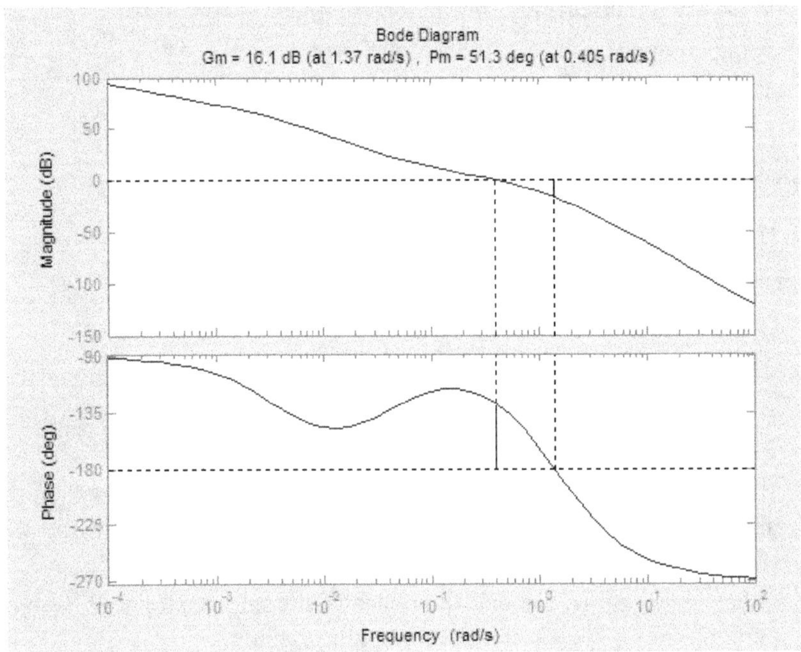

Gain Margin (GM) = 16.1 db

Phase Margin (PM) = 51.3 deg

Phase Cross Over Frequency(wcp) = 1.37 rad/sec

Gain Cross Over Frequency(Wcg) = 0.405 rad/sec

system is stable

Result:

Viva-Voce:

1. What is meant by compensator?

2. What is lag compensator?
3. What are the effects and limitations of phase - lag control.
4 A lag compensator acts like which Filter.
5 To improve which response lag compensator preferred.

6.11 Lead Compensator

Aim: To design a lead compensator for a closed loop system

Apparatus: Personal Computer (PC)

MATLAB Software:

Theory: A lead compensator can be thought of in several different ways.

- First, a lead compensator is a device that provides phase lead in its' frequency response.
- If the compensator has phase lead - and never a phase lag - then there are implications about where the corner frequencies are in the Bode' plot.
- Other implications are that the phase lead compensator will have only certain types of pole-zero patterns in the s plane.

Lead compensators are sometimes the best controller to use to get a system to do what you want it to do. It's as simple as that. They are an option that you may need if you cannot use anything in the PID family to bring a system's performance within specifications. There's no guarantee that a lead compensator will do the trick, but it is another weapon in the arsenal.

We can manipulate TF, SS, and ZPK models using the arithmetic and model interconnection operations described in Operations on LTI Models and analyze them using the model analysis functions, such as bode and step. FRD models can be manipulated and analyzed in much the same way you analyze the other model types, but analysis is restricted to frequency-domain methods.

Using a variety of design techniques, you can design compensators for systems specified with TF, ZPK, SS, and FRD models. These techniques include root locus analysis, pole placement, LQG optimal control, and frequency domain loop-shaping. For FRD models, you can either: Obtain an identified TF, SS, or ZPK model using system identification techniques. Use frequency-domain analysis techniques.

Other Uses of FRD Models: FRD models are unique model types available in the Control System Toolbox collection of LTI model types, in that they don't have a parametric representation. In addition to the standard operations you may perform on FRD models, you can also use them to: Perform frequency domain analysis on systems with nonlinearities using describing functions. Validate identified models against experimental frequency response data.

MATLAB Program:

```
clc;
clear all;
OS=input('Enter % overshoot');
TP=input('Enter peak time');
KV=input('Enter velocity error');
numg=input('Enter the numerator coefficients');
deng=input('Enter the denominatot coefficients');
G=tf(numg,deng)
bode(G)
margin(G)
k=KV/(dcgain(conv([1,0],numg),deng));
G1=zpk(G*k);
z=(-log(OS/100))/(sqrt(pi^2+log(OS/100)^2));
pmreq=atan(2*z/(sqrt(-2*z^2+sqrt(1+4*z^4))))*(180/pi);
wn=pi/sqrt(1-z^2);
wb=wn*sqrt((1-2*z^2)+sqrt(4*z^4-4*z^2+2));
w=logspace(-2,3,1000);
[mag,ang]=bode(G1,w);
[gm pm wcg wcp]=margin(mag,ang,w);
PC=pmreq-pm+15;
beta=(1-sin(PC*pi/180))/(1+sin(PC*pi/180));
magpc=1/sqrt(beta);
for k=1:length(mag)
    if mag(k)-(1/magpc)<=0
        wmax=w(k);
        break;
```

```
    end;
end
zc=wmax*sqrt(beta);
PC=zc/beta;
kc=1/beta;
GC=tf(kc*[1 zc],[1 PC]);
CGC=G1*GC
margin(CGC);
s=tf([1 0],[0]);
KV=dcgain(S*CGC)
CLGC=feedback(CGC,1);
```

Example: Consider a unity feedback system with open loop transfer function $G(s) = \dfrac{K}{s(s+100)(s+36)}$. It is specified that K_v= 40 sec^{-1}, % overshoot 20% and peak time is 0.1 sec. Design a lead compensator to meet the design.

Output:

Transfer function of Un-compensated System

$$G = \frac{100}{s^3 + 136\,s^2 + 3600\,s}$$

Continuous-time transfer function.

Bode plot for Un-compensated System:

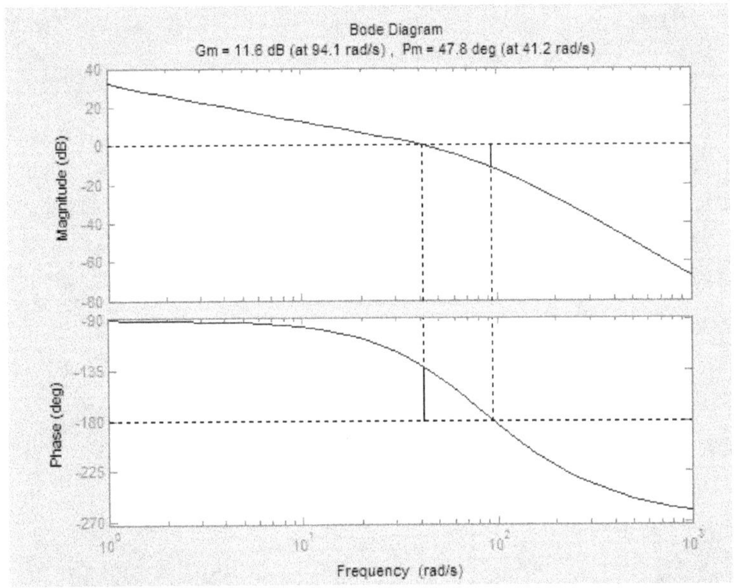

Gain Margin (GM) = 11.6 db

Phase Margin (PM) = 47.8 deg

Phase Cross Over Frequency(wcp) = 94.1 rad/sec

Gain Cross Over Frequency(Wcg) = 41.2 rad/sec

system is stable

Transfer function of compensated System

$$\text{CGC} = \frac{4.1652e05\,(s+24.43)}{s\,(s+100)\,(s+70.67)\,(s+36)}$$

Continuous-time zero/pole/gain model.

Bode plot for Compensated System:

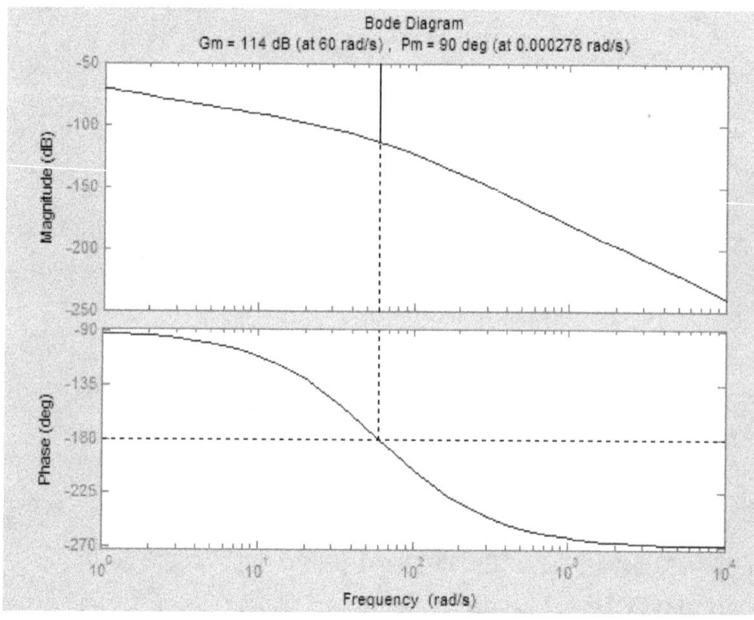

Gain Margin (GM) = 114 db

Phase Margin (PM)= 90 deg

Phase Cross Over Frequency(wcp) = 60 rad/sec

Gain Cross Over Frequency(Wcg) = 0.000278 rad/sec

system is stable

Result:

Viva-Voce:

1. What is meant by compensator?
2. What is lead compensator
3. What are the effects and limitations of phase - lead control?
4. A lead compensator acts like which Filter.
5. How to determine the particular compensator using pole- zero plot.

6.12 Lag- Lead Compensator

Aim : To design lag-lead compensator using closed loop system.

Apparatus: Personal Computer (PC)

MATLAB Software:

Theory: A lead–lag compensator is a component in a control system that improves an undesirable frequency response in a feedback and control system. It is a fundamental building block in classical control theory.

Both lead compensators and lag compensators introduce a pole zero pair into the open loop transfer function. The transfer function can be written in the Laplace domain as

$$\frac{X}{Y} = \frac{s+z}{s+p}$$

where *X* is the input to the compensator, *Y* is the output, *s* is the complex laplace transform variable, *z* is the zero frequency and *p* is the pole frequency. The pole and zero are both typically negative, or left of the zero in the complex plane. In a lead compensator, $| z | < | p |$, while in a lag compensator $| z | > | p |$.

A lead-lag compensator consists of a lead compensator cascaded with a lag compensator. The overall transfer function can be written as

$$\frac{X}{Y} = \frac{(s+z_1)\,(s+z_2)}{(s+p_1)\,(s+p_2)}.$$

Typically $| p_1 | > | z_1 | > | z_2 | > p_2 |$, where z_1 and p_1 are the zero and pole of the lead compensator and z_2 and p_2 are the zero and pole of the lag compensator. The lead compensator provides phase lead at high frequencies. This shifts the poles to the left, which enhances the responsiveness and stability of the system. The lag compensator provides phase lag at low frequencies which reduces the steady state error.

The precise locations of the poles and zeros depend on both the desired characteristics of the closed loop response and the characteristics of the system being controlled. However, the pole and zero of the lag compensator should be close together so as not to cause the poles to shift right, which could cause instability or slow convergence. Since their purpose is to affect the low frequency behaviour, they should be near the origin.

lagts delays a financial time series object by a specified time step.

newfts = lagts(oldfts) delays the data series in oldfts by one time series data entry and returns the result in the object newfts. The end will be padded with zeros, by default.

newfts = lagts(oldfts, lagperiod) shifts time series values to the right on an increasing time scale. lagts delays the data series to happen at a later time. lagperiod is the number of lag periods expressed in the frequency of the time series object oldfts. For example, if oldfts is a daily time series, lagperiod is specified in days. lagts pads the data with zeros (default).

newfts = lagts(oldfts, lagperiod, padmode) lets you pad the data with an arbitrary value, NaN, or Inf rather than zeros by setting padmode to the desired value.

leadts advances a financial time series object by a specified time step.

newfts = leadts(oldfts) advances the data series in oldfts by one time series data entry and returns the result in the object newfts. The end will be padded with zeros, by default.

newfts = leadts(oldfts, leadperiod) shifts time series values to the left on an increasing time scale. leadts advances the data series to happen at an earlier time. leadperiod is the number of lead periods expressed in the frequency of the time series object oldfts. For example, if oldfts is a daily time series, leadperiod is specified in days. leadts pads the data with zeros (default).

newfts = leadts(oldfts, leadperiod, padmode) lets you pad the data with an arbitrary value, NaN, or Inf rather than zeros by setting padmode to the desired value.

MATLAB Program:

```
clc;
clear all;
pos=input('Type %OS  ');
Tp=input('Type peak time   ');
Kv=input('Type value of Kv ');
numg=input('enter the numerator')
deng=input('enter the denominator')
G=tf(numg,deng)
bode(G)
margin(G)
```

```
s=tf([1 0],1);
sG=s*G;
sG=minreal(sG);
K=dcgain(Kv/sG);
G=tf(K*numg,deng);
G=zpk(G)
z=(-log(pos/100))/(sqrt(pi^2+log(pos/100)^2));
Pmreq=atan(2*z/(sqrt(-2*z^2+sqrt(1+4*z^4))))*(180/pi);
wn=pi/(Tp*sqrt(1-z^2));
wBW=wn*sqrt((1-2*z^2)+sqrt(4*z^4-4*z^2+2));
wpm=0.8*wBW;
[M,P]=bode(G,wpm);
Pmreqc=Pmreq-(180+P)+5;
beta=(1-sin(Pmreqc*pi/180))/(1+sin(Pmreqc*pi/180));
zclag=wpm/10;
pclag=zclag*beta;
Kclag=beta;
Glag=tf(Kclag*[1 zclag],[1 pclag]);
Glag=zpk(Glag)
zclead=wpm*sqrt(beta);
pclead=zclead/beta;
Kclead=1/beta;
Glead=tf(Kclead*[1 zclead],[1 pclead]);
Glead=zpk(Glead)
Ge=G*Glag*Glead
 bode(Ge)
margin(Ge)
sGe=s*Ge;
sGe=minreal(sGe);
Kv=dcgain(sGe)
```

Example: Given open loop transfer function $G(s) = \dfrac{1}{s(s+1)(0.5s+1)}$ it is designed to compensate the system so that static velocity error constant kv = 5 sec^{-1}. The peak overshoot is 5% and peak time is 0.1sec. Design a lag-lead compensator to meet the requirements.

Output: Transfer function of Un-Compensated system

$$G = \frac{1}{0.5\,s^3 = 1.5\,s^2 + s}$$

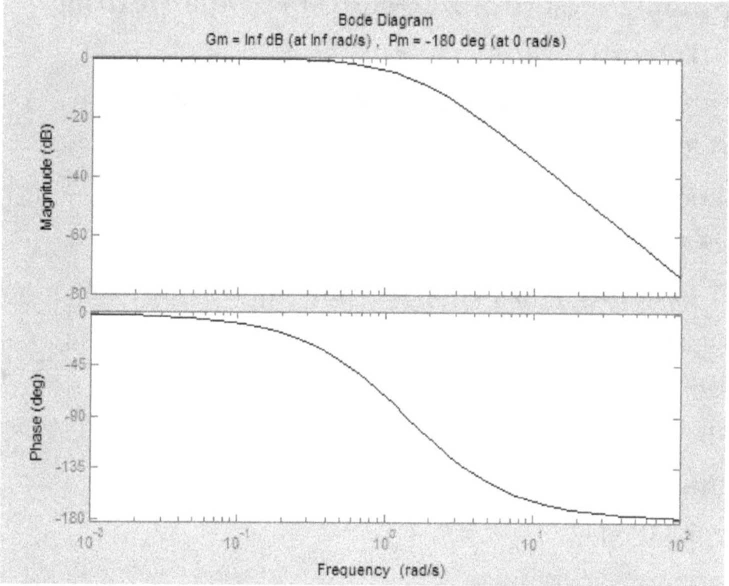

Gain Margin (GM) = infinity db

Phase Margin (PM)= –180 deg

Phase Cross Over Frequency(wcp) = infinity rad/sec

Gain Cross Over Frequency(Wcg) = 0 rad/sec

system is unstable

Transfer function of Compensated system

$$Ge = \frac{10(s+3.556)\,(s+22.57)}{s(s+2)\,(s+1.432)\,(s+1)\,(s+56.04)}$$

Continuous-time zero/pole/gain model.

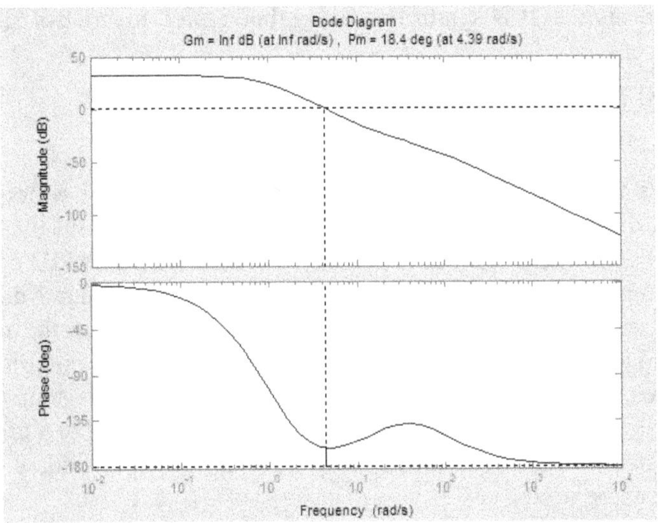

Gain Margin (GM) = Infinity db

Phase Margin (PM) = 18.4 deg

Phase Cross Over Frequency(wcp) = Infinity rad/sec

Gain Cross Over Frequency(Wcg) = 4.39 rad/sec

system is stable

Result:

Viva-Voce:

1. What is meant by compensator?
2. What is lag-lead compensator?
3. What are the types of compensating networks?.
4. How to determine the particular compensator using pole- zero plot.
5. List out merits of Lead-lag network.

6.13 PID Controller Design

Aim: To design a PID controller using bode plot to control the system performance.

Apparatus: Personal Computer (PC)

MATLAB software

Theory: PID controllers are commercially successful and widely used as controllers in industries. For example, in a typical paper mill there may be about 1500 controllers and out of these 90 percent would be PID controllers. The PID controller consists of a proportional mode, an Integral mode and a Derivative mode. The first letters of these modes make up the name PID controller. Depending upon the application one or more combinations of these modes are used. For example, in a liquid control system where we want zero steady state error, a PI controller can be used and in a temperature control system where zero stead state error is not specified, a simple P controller can be used.

The equation of a PID controller in time-domain is given by

$$u(t) = MV(t) = K_p e(t) + K_i \int_0^t e(\tau)\,d\tau + K_d \frac{d}{dt} e(t)$$

where

K_p = Proportional gain, a tuning parameter

K_I = Integral gain, a tuning parameter

K_D = Derivative gain, a tuning parameter

e = Error = *SP-PV*

t = Time or instantaneous time (the present)

τ =Variable of integration; takes on values from time 0 to the present t.

Equivalently, the transfer function in the Laplace Domain of the PID controller is

$L(s) = K_p + K_i/s + K_d\, s$

where s: complex number frequency

The controller used here is a PID controller represented by a block PID and the system or plant is represented by G(s). R(s) and D(s) are reference signal and disturbance signal respectively. Y(s), E(s) and M(s) are the output, error and controller output of the system respectively. For the purpose of good control, we require the system output Y(s) to track any reference signal F(s) and at the same time reject or suppress deviation due to the disturbance signal D(s). Hence the PID controller can realize this objective.

Proportional controller:

MATLAB Program:

```
clc;
clear all;
num=input('enter the numerator of the transfer function')
den=input('enter the denominator of the transfer function') h=tf(num,den)
[gm pm wcp wcg]=margin(h)
km=10*(gm/20)
wm=wcp
kp=0.6*km
ki=(kp*wm)/pi
kd=(kp*ki)/(4*wm)
h1=tf([1,0],[1])
g=(kp+(kd*h1)+(ki/h1))*h
bode(g)
```

Example:

Design a suitable PID controller for the given Transfer function =

$\frac{6}{s(s+5)(s+7)(s+1)}$ using bode plot.

Output:

$$\text{Transfer function(h)} = \frac{6}{s^4 + 13s^3 + 47s^2 + 35s}$$

Continuous-time transfer function.

Gain Margin (GM) = 19.8809

Phase margin (PM) = 77.0994

Phase Cross Over Frequency (wcp) = 1.6408

Gain Cross Over Frequency (wcg) = 0.1689

Proportional Constant kp = 5.9643

Integral Constant ki = 3.1150

Derivative Constant kd = 2.8308

Total Transfer function with controller $g = \dfrac{16.98\,s^2 + 35.79\,s + 18.69}{s^5 + 13\,s^4 + 47\,s^3 + 35\,s^2}$

Continuous-time transfer function.

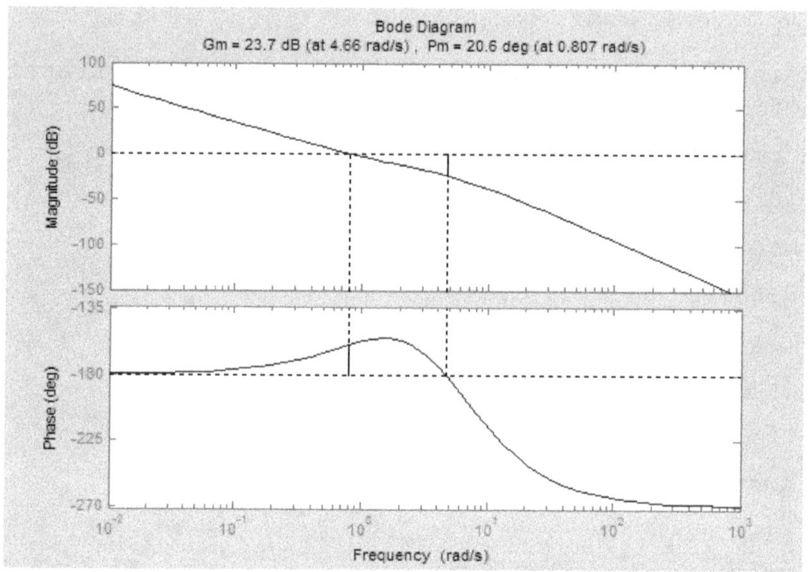

Result:

Viva-Voce :

1. Define proportional?
2. Define integral controller?
3. Why should we do not connect first order& second order in the loop of PID controller?

4. Define second order system?
5. Where shall we apply PID controller?

Exercise Problems

1. For a unity feedback control system with $G(s) = \frac{1}{S(S+1)(0.5S+1)}$, design a phase lag compensator such that the closed loop system will satisfy the following requirements K_v=5/S, phase margin=40^0, gain margin>10db.

2. For the system shown in figure below, design a lead compensator to yield 20% overshoot, K_v= 40 with a peak time of 0.1sec.

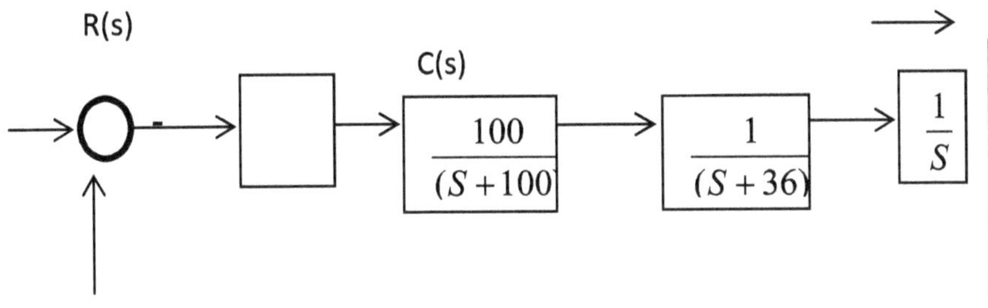

3. For a unity feedback control system with $G(s) = \frac{1}{(S+2)(S+4)}$, design a phase lead compensator such that the closed loop system will satisfy the following requirements, rise time = 0.6sec,% overshoot 15 and settling time 1.66sec.

4. For a unity feedback control system with $G(s) = \frac{1}{S(S+2)}$, design a phase lead compensator such that the closed loop system will satisfy the following requirements, damping ratio 0.5 and natural frequency 4 rad/sec.

7 State Space Analysis in Control Systems

7.1 Introduction

A modern complex system may have many inputs and many outputs, and these may be interrelated in a complicated manner. To analyse such a system, it is essential to reduce the complexity of the mathematical expressions, as well as to resort to computers for most of the tedious computations necessary in the analysis. The state-space approach to system analysis is best suited from this viewpoint.

While conventional control theory is based on the input–output relationship, or transfer function, modern control theory is based on the description of system equations in terms of n first-order differential equations, which may be combined into a first-order vector-matrix differential equation. The use of vector-matrix notation greatly simplifies the mathematical representation of systems of equations. The increase in the number of state variables, the number of inputs, or the number of outputs does not increase the complexity of the equations. In fact, the analysis of complicated multiple-input, multiple-output systems can be carried out by procedures that are only slightly more complicated than those required for the analysis of systems of first-order scalar differential equations. Therefore state space analysis is a very useful technique of analysing control system. It is based on the concept of state and is applicable to linear time varying non-linear and MIMO systems. It employs the use of vector matrix notation. Representation of higher order system becomes very simple. The following are the different definitions come across the system analysis in state space.

7.2 State

The state of control system at time $t=t_0$ is the smallest set of variables called state variables such that the knowledge of the variables along with the knowledge of inputs at $t = t_0$ is sufficient to determine the output dynamics of the system at any time t • t.

In other words, the state of system represents the minimum amount of information needed to know about a system at t_0 such the future behaviour can be determined with reference to input before t_0.

7.3 State Variable

The state variables of a dynamic system are the smallest set of variables that determine the state of the dynamic system, that is the state variables are the minimal set of variables such that the knowledge of these variable at these variables at any initial time $t = t_0$ together with the knowledge of the inputs for t • t_o is sufficient to completely determine the behaviour of the system for any time t • 0.

7.4 State Vector

If n state variable are needed to completely describe the behaviour of a given system then these n state variables can be considered as the n components of a vector x(t). Such a vector is called a state vector. A state vector is thus a vector that determines uniquely the system state x(t) for any time t • t_0, once the state at $t = t_0$ is given and the input u(t) for t • t_0 is specified.

7.5 State Space

If $x_1(t)$, $x_2(t)$....., $x_n(t)$ are the minimum number of state variables which are necessary to describe the dynamics of a control system, then the n - dimensional space whose coordinate axes consist of x_1 axis, x_2 axis,...x_n axis is called a state space. The schematic diagram of state space representation is as shown in Fig.7.1. For the two dimensional cases, the state space reduces to the state plane or phase plane.

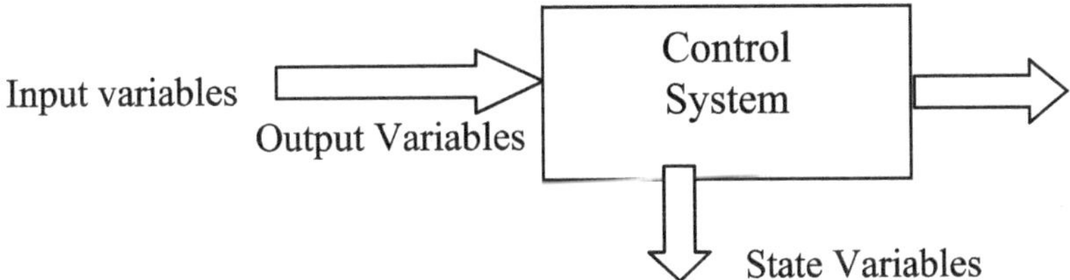

Fig.7.1 Schematic diagram of state space representation

The analysis of control system using state space approach carried out in time domain by representing a system in the form of first order differential equations by selecting suitable state variables such as physical variables, phase variables and canonical variables where in first order derivative terms are arranged on left hand side and right hand side terms are free from derivatives. The variables in such a form of representation are known as state variables. In state space approach MIMO system equations can be arranged in matrix forms which facilitate their solution.

From this analysis the state model can be obtained for electrical, mechanical and electromechanical systems to determine the stability. Therefore the nth order state model of a linear time invariant system is:

$\dot{x}(t)=Ax(t)+Bu(t)$:state equation

$y(t)=Cx(t)+Du(t)$: output equation.

7.6 Stability

The stability of a system represented by its state model since the eigen values of the system matrix are the same as the roots of the characteristic equation which are nothing but poles of the closed loop transfer function. The stability of a system can be determined by determining the location of eigen values. If all the eigen values have only negative real parts then the system representation by the state model is stable. In other hand have positive real parts the state model is unstable. The stability of the system can be determined by finding the controllability and observability of the state model.

7.7 Controllability

The controllability is in relation to transfer of a system from one state to another by appropriate input controls in a finite time. Sometimes such a control not possible and this can be verified by using controllability test matrix. Therefore a system is said to be completely state controllable if any initial state $x(t_0)$ can be transferred to any final state $x(t_f)$ in a finite time t_f• 0by some control input u(t).

The *nxn* controllability matrix is given by

$$R=\left[B\ AB\ A^2B\ \ldots\ A^{n-1}\ B\right]$$

The controllability condition of a system depends on the coefficient matrices A and B and it is said that the pair A, B is controllable indicating that the rank of the test matrix is n. However if controllability matrix is *nxn,* i.e square matrix, then the condition for the state controllability matrix is not equal to zero i.e the matrix is non-singular.

7.8 Observability

The method of determining the state of a system by observing its output concerns observability. Sometimes it is desirable to get information about the state variables from the measurements of the output and input in case it is not possible to obtain any one of the states from the measurement of the output the state is said to be unobservable and the system as a whole is unobservable. The condition under which the state of a system is measured from the output.

A system is said to be completely observable if every state $x(t_0)$ can be exactly determined from measurement of the output y(t) over a finite interval of time t_0 • t• t_f.

$$Q_0 = [C^T : A^T C^T : (A^T)C^T : ....(A^T)^{n-1}C^T]$$

This matrix is to have a rank of n however if the matrix is *nxn* i.e., square matrix then for establishing observability, it should be non-singular. If the observability condition is satisfied i.e., observability test matrix of the rank is then it is said that C, A is observable pair.

Viva-Voce

1. Define the stability in terms of eigen values.
2. Is the state model of a system is unique?
3. What are the two conditions to be satisfied by the state variables?
4. What do you mean by a homogeneous and non-homogeneous state equation?
5. How do you find the order of the system from state space model?

EXPERIMENTS

7.9 Transfer Funtion to State Space Model and Vice Versa

Aim: To convert the given transfer function into State space Model and vice versa.

Apparatus: Personal Computer (PC)

MATLAB software

Theory: The transfer function is defined as the ratio of Laplace transform of output to Laplace transform of input. The transfer function of a given state model is given by:

$$T(s) = c\left[SI - A\right]^{-1} B$$

A state space representation is a mathematical model of a physical system as a set of input, output and state variables related by first-order differential equations. The state space representation (also known as the "time-domain approach") provides a convenient and compact way to model and analyze systems with multiple inputs and outputs. Unlike the frequency domain approach, the use of the state space representation is not limited to systems with linear components and zero initial conditions. "State space" refers to the space whose axes are the state variables. The state of the system can be represented as a vector within that space. The input state equation is given by,

$$\left[\dot{X}\right] = [A][X] + [B][U]$$

The output equation is written as,

$$[Y] = [C][X] + [D][U]$$

There are three methods for obtaining state model from transfer function:

1. Phase variable method
2. Physical variable method
3. Canonical variable method

Out of three methods given above canonical form is probably the most straightforward method for converting from the transfer function of a system to a state space model is to generate a model in "controllable canonical form." This term comes from Control Theory but its exact meaning is not important to us. To see how this method of generating a state space model works, consider the third order differential transfer function. Probably the most straightforward method for converting from the transfer function of a system to a state space model is to generate a model in "controllable canonical form." This term comes from Control Theory but its exact meaning is not important to us. To see how this method of generating a state space model works, consider the third order differential transfer function:

$$H(s) = \frac{Y(S)}{U(s)} = \frac{b_0 s^2 + b_1 s + b_2}{s^3 + a_1 s^2 + a_2 s + a_3}$$

We start by multiplying by Z(s)/Z(s) and then solving for Y(s) and U(s) in terms of Z(s). We also convert back to a differential equation.

$Y(s) = (b_0 s^2 + b_1 s + b_2)\ z(s)$ $\qquad\qquad$ $Y = b_0 z + b_1 z + b_2 z$

$U(s) = (s^3 + a_1 s^2 + a_2 s + a_3)\ z(s)$ $\qquad u = z + a_1 z + a_2 z + a_3 z$

We can now choose z and its first two derivatives as our state variables

$q_1 = z \qquad q_1 = z = q_2$

$q_2 = z \qquad q_2 = z = q_3$

$q_3 = z \qquad q_3 = z = u - a_1 z - a_2 z - a_3 z$

$\qquad = u - a_1 q_3 - a_2 q_2 - a_3 q_1$

Now we just need to form the output

$Y = b_0 z + b_1 z + b_2 z$

$\quad = b_0 q_3 + b_1 q_2 + b_2 q_1$

From these results we can easily form the state space model:

this case, the order of the numerator of the transfer function was less than that of the denominator. If they are equal, the process is somewhat more complex. A result that works in all cases is given below; the details are here. For a general n[th] order transfer function:

$$H(s) = \frac{Y(s)}{U(s)} = \frac{b_0 s^n + b_1 s^{n-1} + \ldots + b_{n-1} s + b_n}{s^n + a_1 s^{n-1} + \ldots + a_{n-1} s + a_n}$$

the controllable canonical state space model form is

$$q = Aq + Bu;\ A = \begin{bmatrix} 0 & 1 & 0 & \cdots & 0 \\ 0 & 0 & 1 & \cdots & 0 \\ \vdots & \vdots & \vdots & \ddots & \vdots \\ 0 & 0 & 0 & \cdots & 1 \\ -a_n & -a_{n-1} & -a_{n-2} & \cdots & -a_1 \end{bmatrix}; \qquad B = \begin{bmatrix} 0 \\ 0 \\ \vdots \\ 0 \\ 1 \end{bmatrix}$$

$Y = Cq + Du \quad c = [b_n - a_n b_0 \quad b_{n-1} - a_{n-1} b_0 \ldots b_2 - a_2 b_0 \quad b_1 - a_1 b_0]\ D = b_0$

MATLAB Program:

```
clc;
clear all;
num=input('Enter the coefficients of numerator')
den=input('Enter the coefficients of denominator')
sys=tf(num,den)
disp('Transfer function to state space conversion')
[A,B,C,D]=tf2ss(num,den)
```

```
disp('state space to transfer function')
[num1,den1]=ss2tf(A,B,C,D)
```

Example: Find state space model for the given transfer function

$$T(s) = \frac{s^3 + 5s^2 + 4s + 6}{4s^3 + 7s^2 + 12s + 9}$$

Output: Transfer function to state space conversion

A = –1.7500 –3.0000 –2.2500

1.0000 0 0

0 1.0000 0

B = 1

0

0

C = 0.8125 0.2500 0.9375

D = 0.250

state space to transfer function

num1 = 0.2500 1.2500 1.0000 1.5000

den3 = 1.0000 1.7500 3.0000 2.2500

Result

Viva-Voce

1. Define transfer function for state space model?
2. What is Zero initial condition for a system means?
3. Define state vector?
4. What is state transition matrix and explain its importance.
5. Mention advantages of state space.

7.10 Controllability

Aim: To determine Controllability of the given system.

Apparatus: Personal Computer (PC)

MATLAB software

Theory: Controllability is an important property of a control system and the controllability property plays a crucial role in many control problems, such as stabilization of unstable system by feedback, or optimal control.

Consider the continuous linear time-invariant system

$$X(t) = Ax(t) + Bu(t)$$

$$Y(t) = Cx(t) + Du(t)$$

where

X is the *nx1* "state vector",

Y is the *mx1* "output vector",

u is the *rx1* "input (or control) vector",

A is the *nxn* "state matrix",

B is the *nxr* "input matrix",

C is the *mxr* "output matrix",

D is the *mxn* "feedthrough (or feedforward) matrix".

The *nxn* controllability matrix is given by

$$R = [B \quad AB \quad A^2B \quad \ldots \quad A^{n-1}B]$$

The system is controllable if the controllability matrix has full row rank (i.e., rank (R) = n).

MATLAB Program:

```
clc;
clear all
disp('program for controllability ');
a=input('Enter matrix a of order(n*n):');
b=input('Enter matrix b of order(n*1):');
s=ctrb(a,b)
```

```
r=rank(s);
if r==size(a)
    disp('System is controllable')
else
    disp('System is uncontrollable')
end
```

Example: Test the controllability of the following state space model.

$$A = \begin{bmatrix} 1.7500 & -3.0000 & -2.2500 \\ 1.0000 & 0 & 0 \\ 0.000 & 1.0000 & 0 \end{bmatrix}$$

$$B = \begin{bmatrix} 1 & 0 & 0 \end{bmatrix}$$

Output:

$$s = \begin{bmatrix} 1.0000 & -1.7500 & 0.0625 \\ 0 & 1.0000 & -1.7500 \\ 0 & 0 & 1.0000 \end{bmatrix}$$

rank = 3

System is controllable

Result:

Viva-Voce:

1. Define controllability.
2. Define State and state variables?
3. What is significance of controllability?
4. State the condition for controllability.
5. When do you say that the system is completely state controllable?

7.11 Observability

Aim: To determine Observability of the given system.

Apparatus: Personal Computer (PC)

MATLAB software

Theory: In control theory **observability** is a measure for how well internal states of a system can be inferred by knowledge of its external outputs. The observability and controllability of a system are mathematical duals. The concept of observability was introduced by American-Hungarian engineer Rudolf e.Kalman for linear dynamic systems.

A system is said to be **observable** if, for any possible sequence of state and control vectors, the current state can be determined in finite time using only the outputs. Less formally, this means that from the systems outputs it is possible to determine the behaviour of the entire system. If a system is not observable, this means the current values of some of its states cannot be determined through output sensors. This implies that their value is unknown to the controller (although they can be estimated through various means).

For time invariant linear systems in the state space representation, there is a convenient test to check if a system is observable. Consider a SISO system with n states if the row ranks of the following *observability matrix*

$$\bullet = \begin{bmatrix} C \\ CA \\ CA^2 \\ \vdots \\ CA^{n-1} \end{bmatrix}$$

is equal to n, then the system is observable. The rationale for this test is that if n rows are linearly independent, then each of the n states is viewable through linear combinations of the output variables $y(k)$.

MATLAB Program:

```
clc;
clear all
disp('program for observability');
a=input('Enter matrix a of order(n*1):');
b=input('Enter matrix b of order(1*n):');
s=obsv(a,b);
r=rank(s);
if r==size(a)
    disp('System is observable')
else
```

```
    disp('System is unobservable')
end
```

Example: Test the observability of the following state space model.

$$A = \begin{bmatrix} 1.0000 & 0 & 0 \\ 0.000 & 1.0000 & 0 \end{bmatrix}$$

$$C = [0.8125 \quad 0.2500 \quad 0.9375]$$

Output:

$$s = \begin{bmatrix} 0.8125 & 0.2500 & 0.9375 \\ -1.1719 & -1.6875 & 2.6367 \\ 0.5508 & 1.6875 & 2.6367 \end{bmatrix}$$

rank = 3

System is observable.

Result:

Viva-Voce:

1. Define observability.
2. Define State Space?
3. What is state observer?
4. What is the significance of observability?.
5. State the condition for observability.

7.12 Step Response of a State Space Model

Aim: To find the step response of a state model for a given system using MATLAB.

Apparatus: Personal Computer

MATLAB Software

Theory: A step signal is a signal whose value changes from one level to another level in zero time. Mathematically, the step signal is represented as given below:

R(t) = u(t) where

$$r(t) = \begin{Bmatrix} 0, & t < 0 \\ 1, & t \geq 0 \end{Bmatrix}$$

In the Laplace transform form,

$$R(s) = \frac{1}{s}$$

The step response of the given transfer function is obtained as follows:

1. The state model is first converted into a transfer function.
2. Step response for that transfer function is obtained as follows.

$$T(s) = \frac{C(s)}{R(s)}$$

and $C(S) = T(S)*R(S)$

$$C(S) = \frac{T(s)}{s}$$

Therefore the output is given by:

$$C(t) = L^{-1}[C(s)]$$

MATLAB Program:

```
clc;
clear all;
A=input('enter matrix A')
B=input('enter matrix B')
C=input('enter matrix C')
D=input('enter matrix D')
Step(A,B,C,D)
```

Example:

Obtain step response of the following state space model.

$$A = \begin{bmatrix} -4 & -16 \\ 1 & 0 \end{bmatrix}$$

$$B = \begin{bmatrix} 1 & 0 \end{bmatrix}$$

$$C = \begin{bmatrix} 0 & 4 \end{bmatrix}$$

$$D = 0$$

Output:

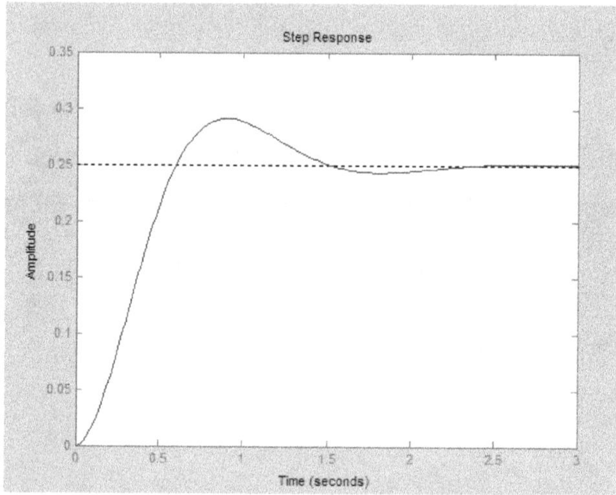

Result:

Viva-Voce:

1. List the applications of state space analysis.
2. Significance of A B C D matrix in state space and their names.
3. Relationship between state equation and transfer function.
4. Define state, state space.
5. Define state vector and state variables.

7.13 Impulse Response of a State Space Model

Aim: To obtain the impulse response of a state model for a given system.

Apparatus: Personal Computer (PC)

MATLAB Software

Theory: In the time domain, we g enerally denote the input to a sy stem as x(t), and the output of the system as y(t). The relationship between the input and t he out put i s de noted a s t he i mpulse r esponse, h(t). W e de fine t he impulse response as b eing the relationship between the system output to it's input. We can use the following equation to define the impulse response:

$$h(t) = \frac{y(t)}{x(t)}$$

Impulse Function It would be handy at this point to define precisely what an " impulse" i s. T he I mpulse F unction, d enoted with •(t) i s a sp ecial function defined piece-wise as follows:

r(t) = •(t) where

$$r(t) = \begin{cases} 0, & t < 0 \\ \text{otherwise} & t = 0 \\ 0, & t > 0 \end{cases}$$

The step response of the gi ven transfer. An examination of the impulse function will show that it is related to the unit-step function as follows:

$$\delta(t) = \frac{du(t)}{dt}$$

and

$$u(t) = \int \delta(t)dt$$

The i mpulse f unction i s not de fined at poi nt t = 0, but t he i mpulse response must always satisfy the following condition, or else it is not a true impulse function:

$$\int_{-\infty}^{\infty} \delta(t)dt = 1$$

The r esponse o f a sy stem t o an i mpulse i nput i s cal led t he i mpulse response. No w, to g et the L aplace T ransform of the i mpulse function, w e take the derivative of the unit step function.

$$L[u(t)] = U(s) = \frac{1}{s}$$

$$L[\delta(t)] = sU(s) = \frac{s}{s} = 1$$

MATLAB Program:

```
Clc;
Clear all;
A=input('enter matrix A')
B=input('enter matrix B')
C=input('enter matrix C')
D=input('enter matrix D')
impulse(A,B,C,D)
```

Example: Obtain the Ramp response of the following state space model.

A = $\begin{bmatrix} -4 & -16 \\ 1 & 0 \end{bmatrix}$

B = $\begin{bmatrix} 1 & 0 \end{bmatrix}$

C = $\begin{bmatrix} 0 & 4 \end{bmatrix}$

D = 0

Output:

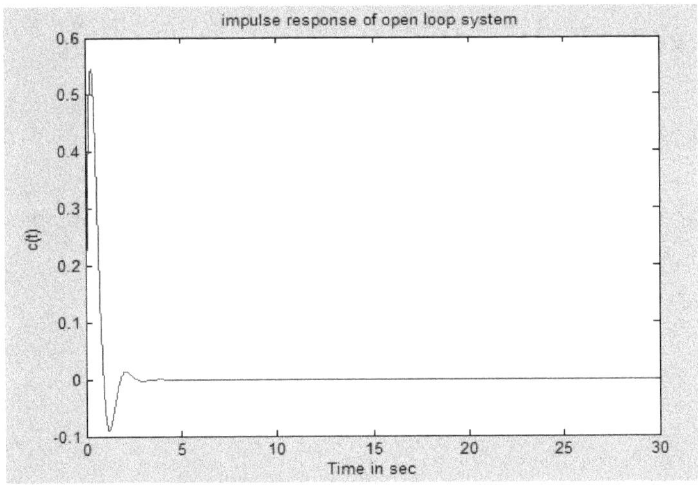

Result:

Viva-voce

1. Define Impulse response
2. Mention the test signals used to determine time response.
3. Define type and order of a system
4. Mention the significance of the impulse response.

Exercise Problems

1. For the state model given below obtain the state models (i) Jordan canonical form ii) plot the pole zero plot of the system.

$$X' = \begin{bmatrix} 0 & 1 & 0 \\ -1 & -2 & -3 \\ 0 & 0 & -3 \end{bmatrix} X + \begin{bmatrix} 0 \\ 1 \\ 1 \end{bmatrix} U$$

$$Y = \begin{bmatrix} 4 & 1 & 0 \end{bmatrix}$$

2. If $G(s) = \dfrac{2S^3 + 11S^2 + 20S + 14}{S^3 + 5S^2 + 7S + 3}$ and $H(s) = C(SI - A)^{-1}B + D$

   (a) Show that $\begin{bmatrix} A - \dfrac{BC}{D} & \dfrac{B}{D} & \dfrac{-C}{D} & \dfrac{1}{D} \end{bmatrix}$ is a

   (b) Show that zeros of $C(SI - A)^{-1} B + D$ can be computed as the eigen values of the matrix

   $A–BD^{-1}C$.

8 Simulink Model

8.1 Introduction to SIMULINK

SIMULINK is an interactive environment for modelling analysing and simulating a wide variety of dynamic systems. SIMULINK provides a Graphical User Interface for constructing block diagram models using drag and drop operations.

Simulation algorithms and parameters can be changed in the middle of a simulation with intuitive results thus providing the user with a ready access learning tool for simulating many of the operational problems found in the real world. SIMULINK is particularly useful for studying the effects of non-linearities on the behaviour of the system and as such it is also an ideal research tool the key features of SIMULINK are given below.

8.2 Key Features

- Graphical editor for building and managing hierarchical block diagrams.
- Libraries of predefined blocks for modelling continuous-time and discrete-time systems.
- Simulation engine with fixed-step and variable-step ODE solvers.
- Scopes and data displays for viewing simulation results.
- Project and data management tools for managing model files and data.
- Model analysis tools for refining model architecture and increasing simulation speed.
- MATLAB Function block for importing MATLAB algorithms into models.
- Legacy Code Tool for importing C and C++ code into models.

8.3 Building the Model

Simulink provides a set of predefined blocks that you can combine to create a detailed block diagram of your system. Tools for hierarchical modeling, data management, and subsystem customization enable you to represent even the most complex system concisely and accurately.

8.4 Selecting Blocks

The Simulink Library Browser includes:

- Continuous and discrete dynamics blocks, such as Integration and Unit Delay.
- Algorithmic blocks, such as Sum, Product, and Lookup Table.
- Structural blocks, such as Mux, Switch, and Bus Selector.

You can build customized functions by using these blocks or by incorporating hand-written MATLAB, C, FORTRAN, or Ada code into your model.

Your custom blocks can be stored in their own libraries within the Simulink Library Browser. Simulink add-on products let you incorporate specialized components for aerospace, communications, PID control, control logic, signal processing, video and image processing, and other applications. Add-on products are also available for modeling physical systems with mechanical, electrical, and hydraulic components.

8.5 Building and Editing the Model

You build a model by dragging blocks from the Simulink Library Browser into the Simulink Editor. You then connect these blocks with signal lines to establish mathematical relationships between system components. Graphical formatting tools, such as smart guides and smart signal routing, help you control the appearance of your model as you build it. You can add hierarchy by encapsulating a group of blocks and signals as a subsystem in a single block.

The Simulink Editor gives you complete control over what you see and use within the model. For example, you can add commands and submenus to the editor and context menus. You can also add a custom interface to a subsystem or model by using a mask that hides the subsystem's contents and provides the subsystem with its own icon and parameter dialog box.

8.6 Navigating through the Model Hierarchy

The Explorer bar and Model Browser in Simulink help you navigate your model. The Explorer bar indicates the level of hierarchy that you are currently viewing and lets you move up and down the hierarchy. The Model Browser provides a complete hierarchical tree view of your model, and like the Explorer bar, can be used to move through the levels of hierarchy.

8.7 Managing Signals and Parameters

Simulink models contain both signals and parameters. Signals are time-varying data represented by the lines connecting blocks. Parameters are coefficients that define system dynamics and behavior.

8.8 Simulink helps you Determine the following Signal and Parameter Attributes

- Data type single, double, signed, or unsigned 8-, 16- or 32-bit integers; Boolean; enumeration; or fixed point.
- Dimensions scalar, vector, matrix, N-D, or variable-sized arrays.
- Complexity real or complex values.
- Minimum and maximum range, initial value, and engineering units.

If you choose not to specify data attributes, Simulink determines them automatically via propagation algorithms, and conducts consistency checking to ensure data integrity.

These signal and parameter attributes can be specified either within the model or in a separate data dictionary. You can then use the Model Explorer to organize, view, modify, and add data without navigating through the entire model.

8.9 Simulating the Model

You can simulate the dynamic behavior of your system and view the results as the simulation runs. To ensure simulation speed and accuracy, Simulink provides fixed-step and variable-step ODE solvers, a graphical debugger, and a model profiler.

8.10 Choosing a Solver

Solvers are numerical integration algorithms that compute the system dynamics over time using information contained in the model. Simulink provides solvers to support the simulation of a broad range of systems, including continuous-time (analog), discrete-time (digital), hybrid (mixed-signal), and multirate systems of any size.

These solvers can simulate stiff systems and systems with discontinuities. You can specify simulation options, including the type and properties of the solver, simulation start and stop times, and whether to load or save simulation data. You can also set optimization and diagnostic information. Different combinations of options can be saved with the model.

8.11 Running the Simulation

You can run your simulation interactively from the Simulink Editor or systematically from the MATLAB command line. The following simulation modes are available:

- Normal (the default), which interpretively simulates your model
- Accelerator, which increases simulation performance by creating and executing compiled target code but still provides the flexibility to change model parameters during simulation
- Rapid Accelerator, which can simulate models faster than Accelerator mode by creating an executable that can run outside Simulink on a second processing core

To reduce the time required to run multiple simulations, you can run those simulations in parallel on a multi-core computer or computer cluster.

8.12 Analysing Simulation Results

After running a simulation, you can analyze the simulation results in MATLAB and Simulink. Simulink includes debugging tools to help you understand the simulation behavior.

8.13 Viewing Simulation Results

You can visualize the simulation behavior by viewing signals with the displays and scopes provided in Simulink. You can also view simulation data within the Simulation Data Inspector, where you can compare multiple signals from different simulation runs.

Alternatively, you can build custom HMI displays using MATLAB, or log signals to the MATLAB workspace to view and analyze the data using MATLAB algorithms and visualization tools.

8.14 Debugging the Simulation

Simulink supports debugging with the Simulation Stepper, which lets you step back and forth through your simulation viewing data on scopes or inspecting how and when the system changes states. With the Simulink debugger you can step through a simulation one method at a time and examine the results of executing that method. As the model simulates, you can display information on block states, block inputs and outputs, and block method execution within the Simulink Editor.

Viva-Voce

1. What do you mean by solver?
2. What are the simulation modes are available to simulate the model?
3. List out the key features of SIMULINK.
4. What is the purpose of Simulation Data Inspector?
5. What do you mean by sub-system?

SIMULINK MODEL

Aim: To study the effect of PID-controllers for the given transfer function using SIMULINK

$$\text{model } T(s) = \frac{100}{100S^2 + 1000S + 10000}$$

Apparatus: Personal Computer (PC)

MATLAB Software:

How to create Simulink diagram?

- At the MATLAB prompt, type Simulink, The SIMULINK block library should appear.
- To open an empty window for operating a new simulation, click on the file menu and select new. This will produce an empty Simulink window.
- In the Simulink block library window, double click on the source icon. Grab the relevant icons from the signal source library, other Simulink libraries and put them in your empty Simulink window (just by placing the mouse arrow on an icon and dragging).
- The Simulink window can be saved by using the file pull-down menu, you will be prompted to name your file name which will be saved with a.mdl extension.

- Restart simulation by typing the file name (without the .mdl extension) at the MATLAB prompt.

Step input without controller:

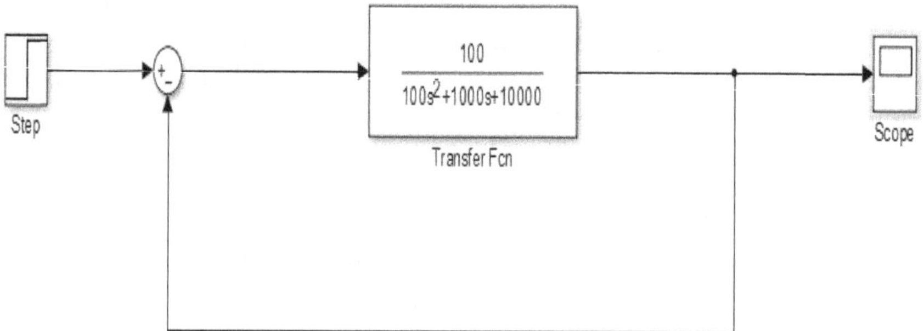

Step Response:

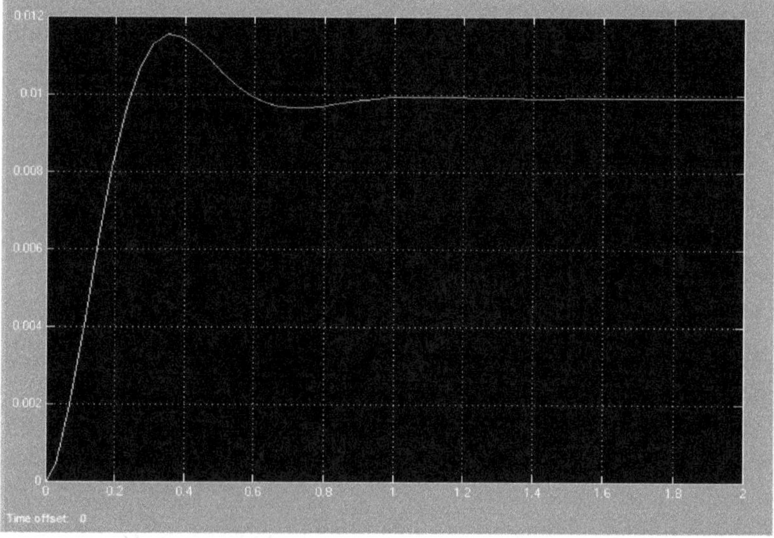

Step input with controller

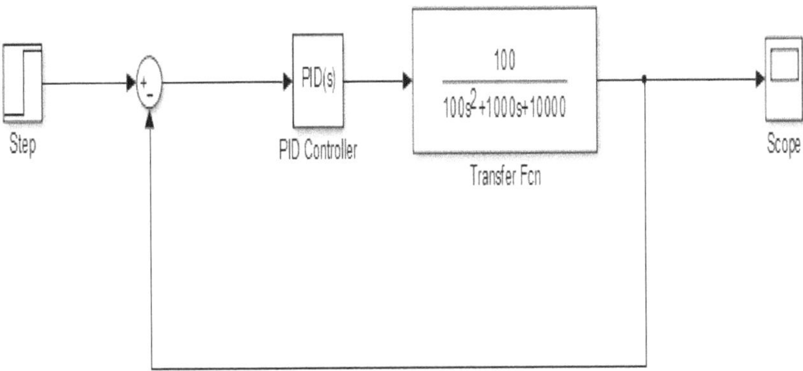

Step Response with PID-Controller:

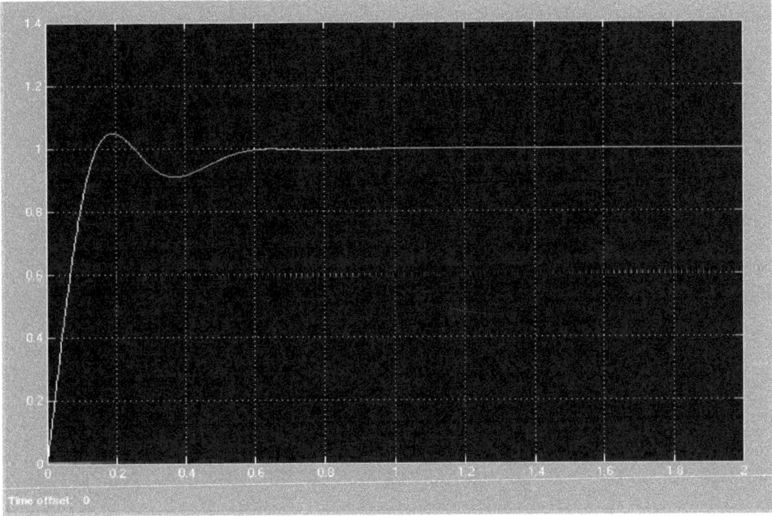

Result:

Viva-Voce:

1. Give the location of scope in Simulink library.
2. How do you tine PID controller?
3. What is the purpose of scope?
4. How do you connect two or more signals to a single input system?
5. Give the location of summing point in Simulink library.

Exercise Problems

1. Obtain the effect of saturation nonlinearity on an open loop system.

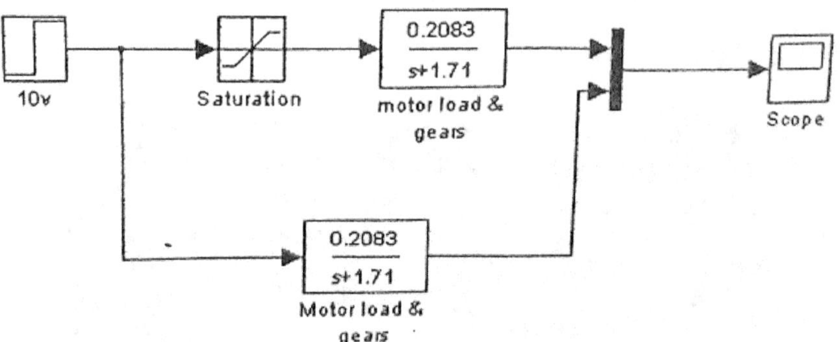

2. Tune the PID controller to obtain the desired response.

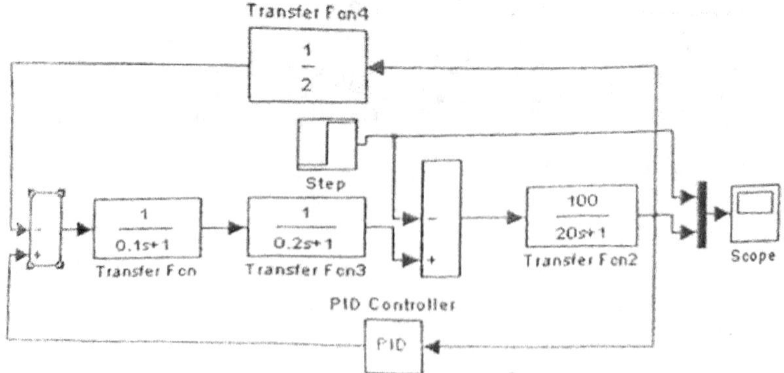

www.ingramcontent.com/pod-product-compliance
Lightning Source LLC
LaVergne TN
LVHW082020150826
845671LV00005B/211

* 9 7 8 9 3 8 6 8 1 9 7 9 6 *